Text Book of Intellectual Property in Agriculture

Text Book of Intellectual Property in Agriculture

Dr. Kalyan Sarma, Ph.D.
Department of Veterinary Medicine
College of Veterinary Sciences and Animal Husbandry
Central Agricultural University
Selesih, Aizawl, (Mizoram)

Dr. Dibyajyoti Talukdar, Ph.D.
Department of Animal Reproduction,
Gynaecology & Obstetrics
College of Veterinary Sciences and Animal Husbandry
Central Agricultural University
Selesih, Aizawl, (Mizoram)

SSPH

SATISH SERIAL PUBLISHING HOUSE
403, Express Tower, Commercial Complex
Azadpur, Delhi-110033 (India)
Phone : 011-27672852, Fax : 91-11-27672046
e-mail : info@satishserial.com, hkjain1975@yahoo.com
Website : www.satishserial.com

Published by :

SATISH SERIAL PUBLISHING HOUSE

403, Express Tower, Commercial Complex, Azadpur, Delhi-110033 (INDIA)
Phone : 011-27672852 Fax : 91-11-27672046
E-mail : info@satishserial.com, hkjain1975@yahoo.com

ISBN 978-93-88020-05-3

Composed, Designed & Printed in India

Preface

Intellectual property rights have never been as much in the news as they are today. With the spread of information technology and new researches being carried out at different levels, the intellectual property protection and the special rights associated with them had now gained special attention. In last few decades with the advancement of globalization, the cross border transactions had increased. Many single national countries companies have now become Multinational Companies (MNC's). But the rights associated with them intellectually are being country specific. Intellectual property rights (IPRs) are specific for the owners and provide the rights to owners such that no one other than person having consent from the owner can use it or define conditions under which it can be used. Here in this book, we will try to give you an idea regarding historical perspectives and need for the introduction of Intellectual Property Right, treaties in IPR, TRIPs and various provision in TRIPs agreement, Indian legislations for the protection of various types of intellectual properties, IPR & biodiversity, copyright, patent, geographical indication, how to file a patent, its future perspective, etc. This book is specially designed for the course PGS-503 for the post-graduate students. We hope, this book will be useful to the students, Scientist as well as the researchers and all other personnel concerned with entrepreneurship, business administration in respect of Intellectual Property Right, various types of intellectual properties, their beneficial advantages, etc.

Kalyan Sarma
Dibyajyoti Talukdar

Contents

Chapter 1

Historical Perspectives and Need for the Introduction of Intellectual Property Rights Regime, the IPR Tools and its Understanding

Introduction

Intellectual property rights have never been as much in the news as they are today. With the spread of information technology and new researches being carried out at different levels, the intellectual property protection and the special rights associated with them had now gained special attention. In last few decades with the advancement of globalization, the cross border transactions had increased. The many single national countries companies have now become Multinational Companies (MNC's). But the rights associated with them intellectually are being country specific. Intellectual property rights (IPRs) are specific for the owners and provide the rights to owners such that no one other than person having consent from the owner can use it or defines conditions, under which it can be used. IPR is characterized by risk in innovation, competition globally, high investment of both money and time and most important human resources needed for the introduction in market. Due to increase in global trade, there has occurred shortening of barriers in the marketing of one product in other country. But the real benefit is to the consumers who are getting the multiple choices for the use of same products. With the advent of trade in goods and services, the IPR has become the most common target of infringement. But the developers are taking care that at least cost of product development could be recovered from the market.

During recent decades, the evolution of developed country IPR regimes has been characterized by three phenomena:

a) The widening of protectable subject-matter. The parameters of protectable subject-matter have been widened, and there has been a tendency to reduce or eliminate exceptions. Examples include the extension of copyright and patent protection to computer programs, the application of patent protection to cover business methods, life-forms, cell lines and DNA sequences, and the removal of exclusions on product patents for drugs.

b) The creation of new rights. Examples of new systems of rights created during the twentieth century include plant breeders' rights, rights to layout-designs of integrated circuits, and rights related to copyright, such as performers' rights.

c) The progressive standardization of the basic features of IPRs. For instance, patent regulations increasingly provide 20-years protection terms; require prior art searches and examinations for novelty, inventive step or non-obviousness, and industrial application; assign rights to the first applicant rather than the first inventor; and provide protection for inventions in all industries and fields of technology.

Difference between intellectual property protection (IPP) and intellectual property rights

Intellectual property (IP) refers to any original creation of the human intellects such as artistic, literary, technical, or scientific creation for which a monopoly is assigned to designated owners by law. Intellectual rights word comes from the French "droits intellectuels".

Intellectual property rights (IPR) refers to the legal rights given to the inventor or creator to protect his invention or creation for a certain period of time. There are certain protection tools designed, such as patent, copyright, trademark, industrials rights design, etc. Such protection is known as Intellectual Property Protection (IPP).

The IPR are the legal rights granted by the government or governing council, which allows the owner of the intellectual property to completely take benefit of the work commercially for a certain period of time.

The Germination stage of the intellectual property system

For the intellectual property rights system, Patent law is the first system in the world. The emergence of the patent system has built an outset to human intellectual property system. At the early in 13th century, the King of Britain had granted a license to the inventors a patent. At the 15th century, the Mediterranean Sea area was filled with the winds of technological innovation in which called on the Governments to establish a new legal system to protect technology.

The Territorial Period	
1311	The birth place of modern intellectual property is **England**, both in patents and copyright. The first recognisable modern form of intellectual property was a Letter Patent, a document bearing the King's seal granting the bearer the right to practise their craft or art in England, the first patent one granted to a Flemish weaver.
1449	The first Letter Patent granted for an invention took place
1474.	The Republic of Venice city constituted the first world's patent law
1623	British patent law came
1641	The first patent in North America was given to Samuel Winslow by the Massachusetts General Court
19th century	Trademarks originated in **Spain**. The trademark system in the modern sense began.
1709	Britain built the first modern copyright law - "the Queen Anne Act.
1791	France adopted its patent law
1793	USA adopted its Patent law
1794	Austria adopted its Patent law
1856	The Act VI of 1856 on protection of inventions based on the British Patent Law of 1852. Certain exclusive privileges granted to inventors of new manufacturers for a period of 14 years.
1857	France set the first legal system to protect trademarks in world
1859	The Act Modified As ACT XV; patent monopolies called exclusive privileges (making, selling and using inventions in India and authorizing others to do so for 14 years from date of filing specification).
1862	Trademark laws first introduced in Britain
1872	The Patents & Designs Protection Act.
1880	The Association Littaeraire et Artistique International was formed in Paris in an attempt to implement some sort of international copyright protection
1883	The Protection of Inventions Act.
1883	Importance of IP first recognized in the Paris Convention for the Protection of Industrial Property
1886	Berne Convention for the Protection of Literary and Artistic Works

1888 Consolidated as the Inventions & Designs Act.

1890 The Sherman Antitrust Act was passed. It basically aims to have the prohibition on the monopoly and give the consumers the varieties to make choices.

1891 Madrid Agreement Concerning the International Registration of Marks

1914 Clayton Antitrust Act was passed. In this law, both substantive and procedural modifications were made in the Sherman antitrust law. In reality, this aimed to confine the anticompetitive practices by beginning the prohibition of particular types of conduct

1911 The Indian Patents & Designs Act.

1925 Hague Agreement Concerning the International Deposit of Industrial Designs

1930 In US the new plant patent act was passed in which the legislation was made to make patent of new varieties of plants, excluding sexual and tuber-propagated plants

1936 Robinson–Patman Act was passed for ruling out some rules of Clayton antitrust law. This rule removed the anticompetitive practices and more important the price discrimination.

1967 One of the most important modern developments in international intellectual property protection is the establishment of the World Intellectual Property Organisation (WIPO).

1970 The Patent Cooperation Treaty

1972 The Patents Act (Act 39 of 1970) came into force on 20TH April, 1972.

1995 The Agreement on Trade Related Aspects of Intellectual Property Rights (TRIPS), which is the most important treaty dealing with developing countries.

1999 On March 26, 1999 Patents (Amendment) ACT, (1999) came into force from 1st January, 1995.

2001 The first patent was published which is a board game designed by Falko Goettsch & James Ernest, and published by Cheapass Games.

2002 The Patents (Amendment) Act, 2002 came into force from 2Oth May, 2003

2005 The Patents (Amendment) Act, 2005 effective from 1st January, 2005

Need for the introduction of Intellectual Property Right regime

The need of most intellectual property law (with the exception of trademarks) is to "Promote progress." By exchanging limited exclusive rights for disclosure of inventions and creative works, society and the patentee/copyright owner mutually benefit, and an incentive is created for inventors and authors to create and disclose their work. The Intellectual Property Right regime helps by :

1. **Financial incentive :** These exclusive rights allow owners of intellectual property to benefit from the property they have created, providing a financial incentive for the creation of an investment in intellectual property, and, in case of patents, pay associated research and development costs.
2. **Economic growth :** The protection of intellectual property rights is essential to maintaining economic growth. It gives statutory expression to the moral and economic rights of creators in their creations and the rights of the public in access to those creations.
3. **Morality :** According to Article 27 of the Universal Declaration of Human Rights, "Everyone has the right to the protection of the moral and material interests resulting from any scientific, literary or artistic production of which he is the author". There are moral arguments for intellectual property. Intellectual property stimulates social progress and pushes people to further innovation.

IPR tools and its understand

There are several tools of intellectual property rights, each distinct from the other in formalities, term of existence and levels of protection. These legal tools are:

a) Patents

b) Trademarks

c) Copyrights and related rights

d) Geographical Indications

e) Industrial Designs

f) Trade Secrets

g) Layout Design for Integrated Circuits

h) Protection of New Plant Variety

Table 1. Illustrates the tools that constitute to the IPR tool kit

IP category	IP protection style
Innovative products and process	Patent, trade secret and utility model
Cultural, artistic and literary work,computer software and compilation data	Copyrights and related rights
Design, including textile design	Industrial design rights
Distinctive sign	Trademarks, collective marks, certification mark, geographical indicators
Microchips	Layout design on topographic of integrated circuits
Denomination for goods of given quality	Geographical indicators
Confidential business and technical information of commerce	Trade secret

It should be realised that each IPR tool protects specific aspects of intellectual creations and it needs to be appreciated that except for copyright (which is governed by the Berne Convention), the rights of the other IPR tools are territorial and therefore, need to be registered in the country in which one desires to have the rights. However, when combined strategically they help to erect formidable protective barriers with strong enforceability.

Role of different tools of IPR are as follows:

Patents	Protect inventions that are novel, non-obvious and useful. Patents have a term of 20 years from the date of filing a complete specification.
Trademark and Service Mark® ™	Protect distinctive marks such as words/signs including personal names,letters, numerals, figurative elements (logos), visually perceptible 2Dor 3D shapes or their combinations capable of distinguishing the goods or services in connection with which it is used in course of trade. Insome countries sounds and distinctive smells can also be registered as trademarks. It can be perpetually renewed from time to time.
Industrial Design Registration	Protect novel non-functional features of shape, configuration, pattern,ornamentation or composition of lines or colors, applied to any article either in two or three-dimensional or in both forms by any industrial process or means whether manual, mechanical or chemical, separate or combined

	which in the finished article appeal to and are judged solely by the eye. This registration has a specific term (initially 10 years and renewable for another term of 10 years).
Copyright ©	Protect creative works that are musical, literary, artistic, lectures, plays,art reproductions, models, photographs, computer software, etc. It is valid for the lifetime of the author and minimum 50 years after the deathof the author.
Layout Designsfor Integrated Circuits	Scope of protection not only includes the protected chip but also the articles incorporating it. The term of protection is 10 years.
Geographical Indications (GI)	Protect the distinctive names of goods that can be identified as originating or manufactured in the territory of a country, or a region or locality in that territory where a given quality, reputation or other characteristic of such goods is essentially attributable to its geographical origin. The term is initially for a period of 10 years and can be renewed perpetually.
Trade Secrets and Undisclosed Information	Protection to persons/institutions on information lawfully under their control from being disclosed to, acquired by or used by others without their consent in a manner contrary to commercial practices so long as.

❑❑❑

Chapter 2

Treaties in IPR, TRIPs and Various Provisions in TRIPs Agreement

Treaties in IPR

Strong protection for intellectual property rights (IPR) worldwide is vital to the future economic growth and development of all countries. Because they create common rules and regulations, international IPR treaties, in turn, are essential to achieving the robust intellectual property protection that spurs global economic expansion and the growth of new technologies.

The international community, however, did not have a single source for intellectual property obligations and norms until the 1994 Uruguay Round of the General Agreement on Tariffs and Trade created the World Trade Organization (WTO) and included the Agreement on Trade-Related Aspects of Intellectual Property Rights (TRIPS).

1. UPOV

- The International Union for the Protection of New Varieties of Plants (UPOV) is an inter-governmental organization with headquarters in Geneva (Switzerland). UPOV was established by the International Convention for the Protection of New Varieties of Plants. The Convention was adopted in Paris in 1961 and it was revised in 1972, 1978 and 1991. The objective of the Convention is the protection of new varieties of plants by an intellectual property right.

2. Trademark Law Treaty

- The Trademark Law Treaty (TLT), adopted on October 27, 1994, entered into force on August 1, 1996.
- Thirty-three states, including the United States, are party to the TLT as of July 1, 2005.

- The TLT was enacted to simplify procedures in the application and registration process and to harmonize trademark procedures in different countries.
- The TLT gives service marks - the distinctive identifiers of businesses that offer a service, as opposed to goods - "equal" status with trademarks. Previously, many foreign countries treated trademarks and service marks differently.
- The TLT requires member nations to register service marks and treat them, as they would trademarks.
- Benefits trademark owners are its prohibition of requirements by national offices for authentication or certification of documents as well as signatures on trademark applications and correspondence.
- Advantage of the TLT is the harmonization of the initial and renewal terms of trademark registration among signatory countries.
- The TLT provides for an initial 10-years term, with 10-years renewals.
- Overall, the TLT is intended to facilitate international trade to individuals and small businesses looking for markets in other countries.

3. Patent Law Treaty

- The Patent Law Treaty (PLT), adopted by WIPO in June of 2000, entered into force on April 28, 2005.
- The PLT is the product of several years of multilateral negotiations on harmonizing global patent systems.
- The PLT harmonizes certain patent application procedures in order to reduce or eliminate formalities and the potential for loss of rights.
- The PLT will make it easier for patent applicants and patent owners to obtain and maintain patents throughout the world by simplifying and, to a large degree, merging national and international formal requirements associated with patent applications and patents.

Budapest treaty

Budapest Treaty on the International Recognition of the Deposit of Micro-organisms for the Purposes of Patent Procedure done at Budapest on April 28, 1977, and amended on September 26, 1980.

1. Rome Convention : Rome Convention for the Protection of Performers, Producers of Phonograms and Broadcasting Organizations
2. Berne Convention: Berne Convention for the Protection of Literary and Artistic Works Paris Convention : Paris Convention for the Protection of Industrial Property

The significance of the TRIPS Agreement is three-fold

1. It is the first single, truly international agreement that establishes minimum standards of protection for several forms of intellectual property;
2. It is the first international intellectual property agreement that mandates detailed civil, criminal, and border enforcement provisions;
3. It is the first international intellectual property agreement that is subject to binding, enforceable dispute settlement.

TRIPS

The Agreement on **Trade-Related Aspects of Intellectual Property Rights** (**TRIPS**) is an international legal agreement between all the member nations of the World Trade Organization (WTO). The TRIPS Agreement came into force in 1995, as part of the Agreement Establishing the World Trade Organization. TRIPS incorporates and builds upon the latest versions of the primary intellectual property agreements administered by the World Intellectual Property Organization (WIPO), the Paris Convention for the Protection of Industrial Property, and the Berne Convention for the Protection of Literary and Artistic Works, agreements that go back to the 1880s.

A TRIP is unique among these IPR accords because membership in the WTO is a "package deal," meaning that WTO members are not free to pick and choose among agreements. They are subject to all the WTO's multilateral agreements, including TRIPS.

TRIPS apply basic international trade principles to member states regarding intellectual property, including national treatment and most-favoured-nation treatment. TRIPS establish minimum standards for the availability, scope, and use of seven forms of intellectual property: copyrights, trademarks, geographical indications, industrial designs, patents, layout designs for integrated circuits, and undisclosed information (trade secrets). It spells out permissible limitations and exceptions in order to balance the interests of intellectual property with interests in other areas, such as public health and economic development.

According to TRIPS, developed countries were to have implemented the agreement fully by January 1, 1996. Developing-country members and members in transition to a market economy were entitled to delay full implementation of TRIPS obligations until January 1, 2000. Least-developed members were given until January 1, 2006, to implement their obligations, with the possibility of further transition upon request. Developing countries that did not provide patent protection for particular areas of technology on their date of application were given an additional five years, until January 1, 2005, to provide such protection.

At the 2001 WTO Ministerial Conference in Doha, least-developed countries were given an additional 10 years to implement TRIPS patent and "undisclosed information" provisions, as they relate to pharmaceuticals.

Because the TRIPS Agreement is a decade old, however, it does not address several new developments, such as the Internet and digital copyright issues, advanced biotechnology, and international harmonization, the process of creating uniform global standards of laws or practice. It sets the floor for minimum IPR protection, not the ceiling.

Since the conclusion of the TRIPS Agreement, the World Intellectual Property Organization has addressed digital copyright issues in the so-called Internet Treaties, namely the WIPO Copyright Treaty (WCT) and the WIPO Performances and Phonograms Treaty (WPPT).

The requirement of TRIPS

TRIPS require member states to provide strong protection for intellectual property rights. For example, under TRIPS:

- Copyright terms must extend at least 50 years, unless based on the life of the author (Art. 12 and 14).
- Copyright must be granted automatically, and not based upon any "formality," such as registrations, as specified in the Berne Convention (Art. 9).
- Computer programs must be regarded as "literary works" under copyright law and receive the same terms of protection.
- National exceptions to copyright (such as "fair use" in the United States) are constrained by the Berne three-step test.
- Patents must be granted for "inventions" in all "fields of technology" provided they meet all other patentability requirements (although exceptions for certain public interests are allowed (Art. 27.2 and 27.3) and must be enforceable for at least 20 years (Art. 33).
- Exceptions to exclusive rights must be limited, provided that a normal exploitation of the work (Art. 13) and normal exploitation of the patent (Art. 30) is not in conflict.
- No unreasonable prejudice to the legitimate interests of the right holders of computer programs and patents is allowed.
- Legitimate interests of third parties have to be taken into account by patent rights (Art. 30).
- In each state, intellectual property laws may not offer any benefits to local citizens, which are not available to citizens of other TRIPS signatories under the principle of national treatment (with certain limited exceptions, Art. 3 and 5). TRIPS also have a most favoured nation clause.

Madrid System for the International Registration of Marks

The Protocol Relating to the Madrid Agreement Concerning the International Registration of Marks – the Madrid Protocol – was adopted in Spain's capital on June 27, 1989, and entered into force on December 1, 1995. The protocol is one of the two treaties comprising the Madrid System for international registration of trademarks. The first treaty, the 1891 Madrid Agreement, provides for the registration of trademarks in several countries through the filing of one international trademark registration with WIPO in Geneva.

The Madrid Protocol, developed because some countries had problems with the operation of the Madrid Agreement, is seen as an improvement to the system for international registration of trademarks. As a result, more and more trademark owners are using the Madrid Protocol every year to protect their trademarks in foreign countries. As of September 15, 2005, there were 66 contracting parties to the Madrid Protocol.

The Madrid Protocol is a filing treaty and not a substantive harmonization treaty. It provides a cost-effective and efficient way for trademark holders - individuals and businesses - to ensure protection for their marks in multiple countries through the filing of one application with a single office, in one language, with one set of fees, in one currency. Moreover, no local agent is needed to file the application. Applications may be filed in English, French, or Spanish.

An application for international registration has the same effect as a national application for registration of the mark in each of the countries designated by the applicant. Once the trademark office in a designated country grants protection, the mark is protected just as if that office had registered it.

The Madrid Protocol also simplifies the subsequent management of the mark, since a simple, single procedural step serves to record subsequent changes in ownership or in the name or address of the holder with WIPO's International Bureau.

Before the protocol was enacted, burdensome administrative requirements for the normal transfer of business assets often made it difficult for trademark owners to carry out valid assignments of their marks internationally. The protocol allows the holder of an international registration to file a single request with a single payment, in order to record the assignment of a trademark with all the member countries. Registration renewal also involves a simple, single procedural step. International registration lasts 10 years, with 10-years renewal periods.

Trademark owners may designate additional countries if they decide to seek protection in more member countries or if new countries accede to the protocol.

If the basic application - or registration upon which the international registration is based, is cancelled for any reason in the first five years, the Madrid Protocol gives the holder of the international registration the opportunity to turn the international registration into a series of national applications in each designated country. This series of applications keeps the priority date of the original international registration in each country. The holder also preserves the rights acquired in each member country, even if international registration fails.

The Hague system for the international deposit of industrial designs

The Hague System is an international registration system that enables owners to obtain protection for their industrial designs with a minimum of formality and expense. A single international application filed with WIPO's International Bureau replaces a whole series of applications previously required in a number of states and/or inter-governmental organizations party to the Hague System. The subsequent management of the international registration is considerably easier under this system. For example, one single step is all that is needed to record a change in the name or address of the holder, or a change in ownership for some or for all of the designated contracting parties.

The Hague System had 42 contracting parties as of April 26, 2005.

Budapest treaty on the international recognition of the deposit of micro-organisms for the purpose of patent procedure

The Budapest Treaty on the International Recognition of the Deposit of Micro-organisms for the Purpose of Patent Procedure, signed on April 28, 1977, was amended on September 26, 1980. The Budapest Treaty eliminates the need to deposit micro-organisms in each country, where patent protection is sought.

Under the treaty, the deposit of a micro-organism with an "international depositary authority" satisfies the deposit requirements of treaty members' national patent laws. An "international depositary authority" is capable of storing biological material and has established procedures that assure compliance with the Budapest Treaty. Such procedures include requirements that the deposit will remain available for the life of the patent and that samples will be furnished only to those persons or entities entitled to receive them.

The establishment of "international depositary authorities" offers several advantages to both patent applicants and contracting states. Patent applicants benefit because the need to deposit in many countries in which they seek patent protection is dramatically reduced. Since a single deposit in any "international depositary authority" will satisfy the national disclosure requirements of any member state, patent applicants' costs are much lower. Using a single authority as a deposit increases the deposit's security, and provides a mechanism of distribution of the deposit. Contracting states benefit because they can rely on the treaty's uniform standards to assure effective deposit and public availability. They no longer need to independently establish a 'recognized' depositary to meet national patentability disclosure requirements.

As of May 2005, there are 60 Patent Offices that abide by the terms of the Budapest Treaty and 35 "international depositary authorities" in 22 different countries.

International Convention for the Protection of New Varieties of Plants

The International Convention for the Protection of New Varieties of Plants (UPOV) established an internationally recognized intellectual property system for the protection of new plant varieties. The UPOV Convention encourages and rewards the ingenuity and creativeness of breeders developing new varieties of plants. Anyone who develops a new variety of plant that may be disease resistant, drought resistant, cold tolerant, or simply aesthetically more pleasing is no less an inventor than someone who improves an automobile engine or develops a new medicinal drug. The only difference is that the plant breeder works with living material, rather than inanimate matter.

The process of creating a new plant variety is often long and expensive. Reproducing an existing plant variety, however, can be quick and relatively easy. Thus, an effective system of intellectual property protection needs to reward innovation by permitting inventors to recover their investment and, at the same time, disseminate the knowledge of that innovation for others to improve upon. The UPOV system establishes basic legal principles of protection that reward breeders for their inventiveness by providing exclusive rights to their plant invention, while encouraging the development of new plant varieties.

Under the 1991 UPOV system, the most recently concluded of these, the exclusive rights granted to the inventor (commonly referred to as "breeder's rights") require that another party other than the owner of the breeder's rights receive the breeder's authorization to:

- Produce or reproduce the protected variety;
- Condition the variety for propagation purposes and
- Offer to sell or market, import, export, or to stock the protected variety.

To receive a breeder's right, a breeder must invent a plant variety that is new, distinct, uniform, and stable. Under the UPOV Convention, however, a plant breeder generally does not need breeder authorization to use protected plant varieties for non-commercial or experimental acts or acts done for the purposes of breeding new plant varieties. The UPOV Convention also allows each member nation to restrict the breeder's right in relation to any variety to allow farmers to use part of their harvest for subsequent plantings in their own land. These restrictions, however, must be within reasonable limits and subject to the safeguarding of the legitimate interests of the breeder.

UPOV member states hold biannual meetings of the Council, a permanent body of the convention. Other UPOV bodies include the Consultative Committee, the Administrative and Legal Committee, and the Technical Committee, made up of several Technical Working Parties (TWPs) across several agricultural sectors. The TWPs meet periodically to share and discuss observations and advancements in agricultural sectors, which helps to standardize examination standards among member states. These TWP meetings benefit breeders as well, since more uniform standards lead to greater consistency of application filings in different territories.

As of June 29, 2005, there were 59 member States to the UPOV Convention. UPOV membership is expected to continue to increase in the next several years.

Conclusions

In the information age, with technology advancing at an accelerating rate, simply implementing the TRIPS Agreement is not enough to establish a robust intellectual property system. While it was the first comprehensive IPR agreement of its time, it is a decade old, and reflects a "snapshot" in time. Technological advances in information technology, biotechnology, and other fields require the updating of national and international laws that protect IP. Fortunately, WIPO has led the way in developing new international norms to meet these challenges.

WIPO also has led the way in simplifying and streamlining the procedures for seeking, obtaining, and maintaining rights in multiple countries. Through its "Global Protection Services" and its harmonization treaties, it saves creators and national IP offices a great deal of time and effort. WIPO also makes available its excellent technical assistance for establishing and improving IPR systems worldwide. Countries should look to both the WTO and to WIPO, when crafting their IPR systems.

❑❑❑

Chapter 3

Intellectual Property Rights Benefits of Securing IPRS

Introduction

Intellectual property rights are a term used for various legal entitlements which attach to certain types of information, ideas or other intangible in their expressed form. The term intellectual property reflects the idea that this subject-matter is the product of the mind or the intellect and the intellectual property rights may be protected at law in the same way as any other form of property. Intellectual property (IP) pertains to any original creation of the human intellect, such as artistic, literary, technical, or scientific creation. Intellectual property rights (IPR) refers to the legal rights given to the inventor or creator to protect his invention or creation for a certain period of time. These legal rights confer an exclusive right to the inventor/creator or his assignee to fully utilize his invention/creation for a given period of time. It is very well-settled that IP play a vital role in the modern economy.

Types of Intellectual Property Rights

- Patents
- Trademarks
- Copyrights
- Geographical indications
- Industrial designs
- Trade secrets
- Layout designs for integrated circuits
- Protection of new plant variety

Patents

A **patent** is a set of exclusive rights granted by a sovereign state to an inventor or assignee for a limited period of time in exchange for detailed public disclosure of an invention. An invention is a solution to a specific technological problem and is a product or a process. Patents are a form of intellectual property.

The word *patent* originates from the Latin *patere*, which means "to lay open" (i.e., to make available for public inspection). More directly, it is a shortened version of the term *letters patent*, which was an open document or instrument issued by a monarch or government granting exclusive rights to a person, predating the modern patent system. Similar grants included land patents, which were land grants by early state governments in the USA, and printing patents, a precursor of modern copyright.

In modern usage, the term *patent* usually refers to the right granted to anyone who invents any new, useful, and non-obvious process, machine, article of manufacture, or composition of matter. Some other types of intellectual property rights are also called *patents* in some jurisdictions: industrial design rights are called *design patents* in the US, plant breeders' rights are sometimes called *plant patents*, and utility models and *Gebrauchsmuster* are sometimes called *petty patents* or *innovation patents*.

The additional qualification *utility patent* is sometimes used (primarily in the US) to distinguish the primary meaning from these other types of patents. Particular species of patents for inventions include biological patents, business method patents, chemical patents and software patents.

Advantages of getting a patent

- A patent gives you the right to stop others from copying, manufacturing, selling or importing your invention without your permission.
- You get protection for a pre-determined period, allowing you to keep competitors at bay.
- You can then use your invention yourself.
- Alternatively, you can license your patent for others to use it, or sell it, as with any asset. This can provide an important source of

revenue for your business. Indeed, some businesses exist solely to collect the royalties from a patent they have licensed-perhaps in combination with a registered design and trademark.

Disadvantages of getting a patent

- Your patent application means making certain technical information about your invention **publicly available**. It might be that keeping the details of your invention **secret** will keep competitors at bay more effectively.
- Applying for a patent can be a very time-consuming and lengthy process (typically three to four years) - market may have changed or technology may have overtaken your invention by the time a patent is granted.
- Cost - it will cost you money whether you are successful or not - the application, searches for existing patents and a patent attorney's fees can all contribute to a reasonable outlay.
- You'll need to remember to pay your annual fee or your patent will lapse.
- You'll need to be prepared to defend your patent. Taking action against an infringer can be very expensive. On the other hand, a patent can act as a deterrent, making defence unnecessary.

What are the advantages of getting patent in India?

Advantages of owning patent would be:

- You own the invention for given time (20 years)
- You can use it to build a business
- Rent it (in this case license it) to existing businesses
- Exclude all others for using, selling, offering for sale and importing your invention in your country
- You can completely sell the patent to other company

Hence, this gives a **unique advantage to patent owner** which can be leveraged for:

1. Building business around the invention
2. Having patent helps in raising capital for business

3. Patent rights can be licensed to other companies to receive royalty payments
4. You can sell the patent
5. Being owner of the patent you have rights to block others from:
   a) Making,
   b) Using,
   c) Selling or offering for sale, and
   d) Importing the patented invention
6. Thus, businesses and inventors can have a complete monopoly and competitive advantage.

Patentable subject-matter

- Any article, apparatus or machinery or its component
- Any substance whether living or non-living, product, pharmaceutical product

Patentable subject-matter

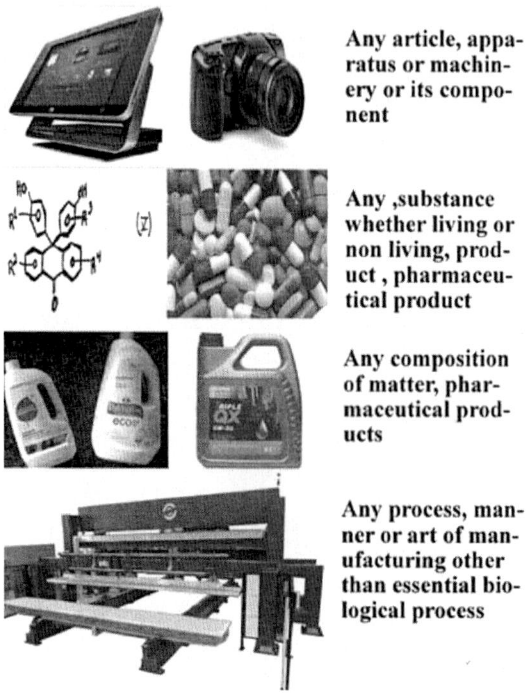

- Any composition of matter, pharmaceutical products
- Any process, manner or art of manufacturing other than essential biological process.

What is not Patentable?

1. Inventions falling within the scope of Sec. (3) of Patents Act, 1970.
2. Inventions falling within the scope of Sec. (1) of Sub-sec. 20 of Atomic Energy Act, 1962- For Example : Inventions relating to Compounds of - Uranium, Beryllium, Thorium, Plutonium, Radium, Graphite, Lithium and more as notified by Central Govt. from time to time.

   Inventions Not Patentable-Sec. 3

3 a) Frivolous or anything obviously contrary to well-established natural laws;

b) Contrary public order or morality serious prejudice to human, animal or plant life or health or to the environment;

c) The mere discovery of a scientific principle or the formulation of an abstract theory or discovery of any living thing or non-living substances occurring in nature;

d) The mere discovery of a new form of a known substance which does not result in the enhancement of the known efficacy of that substance;

e) A substance obtained by a mere admixture resulting only in the aggregation of the properties of the components thereof or a process for producing such substances;

h) A method of agriculture or horticulture;

i) Any process for the medicinal, surgical, curative, prophylactic, diagnostic, therapeutic or other treatment of animals to render them free of disease or to increase their economic value or that of their products;

j) Plants & animals in whole or any part thereof other than micro-organisms but including seeds, varieties and species and essentially biological processes for the production or propagation of plants and animals.

k) A mathematical or business method or a computer programme per se or algorithms

l) A literary, dramatic, musical or artistic work or any other aesthetic creation

m) Mere scheme or rule or method of performing mental act or method of playing game

n) Topography of integrated circuits

p) An invention which in effect is traditional knowledge or which is an aggregation or duplication of known properties or traditionally known component or components.

How can you get a Patent in India?

Step 1: Write down the invention (idea or concept) with as much details as possible

Collect all the information about your invention, such as:

- Area of invention
- Description of the invention what it does?
- How does it work?
- Advantages of the invention

Ideally, if you have worked on the invention during research and development phase, you should have something call lab record duly signed with date by you and respective authority.

Step 2: Include drawings, diagrams or sketches explaining working of invention

The drawings and diagrams should be designed so as to explain the working of the invention in better way with visual illustrations. They play an important role in patent application.

Step 3: Check whether the invention is patentable subject-matter

All inventions may not be patentable, as per Indian patent act, there are certain inventions that are not patentable.

Step 4 a: Patentability search

The next step would be finding out whether your invention meets all patentability criteria as per Indian patent act? That is,

- Novelty
- Non-obviousness
- Industrial application
- Enabling

Step 4 b: Decide whether to go ahead with patent

The patentability report and opinion helps you decide whether to go ahead with the patent or not, chances are what you thought as novel might already been patented or know to public in some form of information. Hence, this reports save lots of time, efforts and cost of the inventor by helping him decide whether to go ahead with the patent filing process or not.

Step 5: Draft (write) patent application

In case you are at very early stage in the research and development for your invention, then you can go for **provisional application**. It gives following benefits:

- Secures filing date
- 12 months of time to file complete specification
- Low cost

After filing provisional application, you secure the filing date which is very crucial in patent world. You get 12 months of time to come up with the complete specification, upon expiry of 12 months your patent application will be abandoned.

When you complete the required documents and your research work is at level, where you can have prototype and experimental results to prove your inventive step you can file complete specification with patent application.

Filing the provisional specification is the optional step, if you are at the stage where you have complete information about your invention, then you can directly go for complete specification.

Step 6: Publication of the application

Upon filing the complete specification along with application for patent, the application is published after 18 months of first filing.

An early publication request can be made along with prescribed fees, if you do not wish to wait till the expiry of 18 months from the date of filing for publishing your patent application.

Generally, the patent application is published within a month form request form early publication.

Step 7: Request for examination

The patent application is examined only after receiving request for examination that is RFE. Upon receiving this request, the controller gives your patent application to a patent examiner who examinees the patent application with different patentability criteria-like:

- Patentable subject-matter
- Novelty
- Non-obviousness
- Inventive step
- Industrial application
- Enabling

The examiner creates a first examination report of the patent application upon reviewing it for above terms. This is called patent prosecution. Everything happening to patent application before grant of patent is generally called as patent prosecution.

The first examination report submitted to controller by examiner generally contains prior arts (existing documents before the date of filing), which are similar to the claimed invention, and same is reported to patent applicant.

Step 8: Respond to objections

Majority of patent applicants will receive some type of objections based on examination report. The best thing to do it, analyse the examination report with patent professional (patent agent) and creating a response to the objections raised in the examination report.

This is a chance for an inventor to communicate his novelty over prior arts found in the examination report. The inventor and patent agent create and send a response to the examination that tries to prove to controller that his invention is indeed patentable and satisfies all patentability criteria's.

Step 9: Clearing all objections

This communication between controller and patent applicant is to ensure that all objections raised in the patent application are resolved. (if not, the patent will not be granted) and the inventor has his fair chance to prove his point and establish novelty and inventive step over existing prior arts. Upon finding the patent application in order of grant, it is grant to the patent applicant as early as possible.

Step 10: Grant of patent

The application would be placed in order for grant, once it is found to be meeting all patentability requirements. The grant of patent is notified in the patent journal, which is published time to time.

Where to file patent application?

The appropriate office of the patent shall be the head office of the patent office or the branch office, as the case may be within whose territorial limits. The patent office is near to the residence of applicant or domicile or his place of business; or the place where the invention actually originated. If the applicant has no business or domicile in India, the address for service is:

Patent Office	Territorial Jurisdiction
Patent Office Branch, Chennai	The States of Andhra Pradesh, Karnataka, Kerala, Tamil Nadu and the Union Territories of Puducherry and Lakshadweep
Patent Office Branch, Mumbai	The States of Maharashtra, Gujarat, Madhya Pradesh, Goa and Chhattisgarh and the Union Territories of Daman and Diu & Dadra and Nagar Haveli.
Patent Office Branch, New Delhi	The States of Haryana, Himachal Pradesh, Jammu and Kashmir, Punjab, Rajasthan, Uttar Pradesh, Uttaranchal, Delhi and the Union Territory of Chandigarh.
Patent Office, HO Kolkata	The rest of India

Trademarks

- A trademark is a sign capable of distinguishing the goods or services of one enterprise from those of other enterprises.
- It protects the public from confusion and deception by identifying the source or origin of products, as distinguished some other similar products.
- It may consist of drawings, symbols, three-dimensional signs, such as the shape and packaging of goods.
- The owner of a registered trademark must be granted the exclusive right to prevent all third parties not having the owner's consent from using in the course of trade identical or similar signs for goods or services, which are identical or similar to those in respect of which the trademark is registered, where such use would result in a likelihood of confusion. In case of the use of an identical sign for identical goods or services, a likelihood of confusion must be presumed.

Usage of Trademarks

A trademark identifies the brand owner of a particular product or service. Trademarks can be licensed to others; for example, Bully land obtained a license to produce Smurf figurines; the Lego Group purchased a license from Lucas film in order to be allowed to launch Lego Star Wars; TT Toys Toys is a manufacturer of licensed ride-on replica cars for children. The unauthorized usage of trademarks by producing and trading counterfeit consumer goods is known as brand piracy.

The owner of a trademark may pursue legal action against trademark infringement. Most countries require formal registration of a trademark as a precondition for pursuing this type of action. The United States, Canada and other countries also recognize common law trademark rights, which means action can be taken to protect an unregistered trademark, if it is in use. Still, common law trademarks offer the holder in general less legal protection than registered trademarks.

A trademark may be designated by the following symbols:

- ™ (the "trademark symbol", which is the letters "TM" in superscript, for an unregistered trademark, a mark used to promote or brand goods)

- SM (which is the letters "SM" in superscript, for an unregistered service mark, a mark used to promote or brand services)
- ® (the letter "R" surrounded by a circle, for a registered trademark).

A trademark is typically a name, word, phrase, logo, symbol, design, image, or a combination of these elements. There is also a range of non-conventional trademarks comprising marks which do not fall into these standard categories, such as those based on colour, smell, or sound (like jingles). Trademarks which are considered offensive are often rejected according to a nation's trademark law.

The term *trademark* is also used informally to refer to any distinguishing attribute by which an individual is readily identified, such as the well-known characteristics of celebrities. When a trademark is used in relation to services rather than products, it may sometimes be called a service mark, particularly in the United States.

Copyrights

Copyright is a legal right created by the law of a country that grants the creator of an original work exclusive rights for its use and distribution. This is usually only for a limited time. The exclusive rights are not absolute but limited by limitations and exceptions to copyright law, including fair use. A major limitation on copyright is that copyright protects only the original expression of ideas, and not the underlying ideas themselves.

Copyright is a form of intellectual property, applicable to certain forms of creative work. Some, but not all jurisdictions require "fixing" copyrighted works in a tangible form. It is often shared among multiple authors, each of whom holds a set of rights to use or license the work, and who are commonly referred to as rights holders. These rights frequently include reproduction, control over derivative works, distribution, public performance, and moral rights, such as attribution.

Copyrights are considered "territorial rights", which means that they do not extend beyond the territory of a specific jurisdiction. While many aspects of national copyright laws have been standardized through international copyright agreements, copyright laws vary by country.

Typically, the duration of a copyright spans the author's life plus 50 to 100 years (that is, copyright typically expires 50 to 100 years after the author dies, depending on the jurisdiction). Some countries require certain copyright formalities to establishing copyright, but most recognize copyright in any completed work, without formal registration. Generally, copyright is enforced as a civil matter, though some jurisdictions do apply criminal sanctions.

Most jurisdictions recognize copyright limitations, allowing "fair" exceptions to the creator's exclusivity of copyright and giving users certain rights. The development of digital media and computer network technologies have prompted re-interpretation of these exceptions, introduced new difficulties in enforcing copyright, and inspired additional challenges to copyright law's philosophic basis. Simultaneously, businesses with great economic dependence upon copyright, such as those in the music business, have advocated the extension and expansion of copyright and sought additional legal and technological enforcement.

Limitation of copyrights

In many jurisdictions, copyright law makes exceptions to these restrictions, when the work is copied for the purpose of commentary or other related uses. It should be noted that US copyright does NOT cover names, title, short phrases or Listings (such as ingredients, recipes, labels, or formulas). However, there are protections available for those areas copyright does not cover - such as trademarks and patents.

There are some exceptions to what copyright will protect. Copyright will not protect:

- Names of products
- Names of businesses, organizations, or groups
- Pseudonyms of individuals
- Titles of works
- Catchwords, catchphrases, mottoes, slogans, or short advertising expressions
- Listings of ingredients in recipes, labels, and formulas, though the directions can be copyrighted

Terms of copyright in India

S.No.	Type of work	Term
1.	Literary work (within life time)	Lifetime + 60 years
2.	Cinematography, films, records, photographs, posthumous publications, anonymous publication, works of government & international agencies	60 years (next to the year of publication)
3.	Broadcasting	25 years (-do-)

Geographical Indications

- Indications which identify a good as originating in the territory of a member, or a region or a locality in that territory, where a given quality, reputation or characteristic of the good is essentially attributable to its geographic location.
- More importantly, it identifies the product's special characteristics, which are the result of the product's origin. Examples :
  - Darjeeling tea
  - In Rajasthan- Bikaneri Bhujia.
  - In Andhra Pradesh- Tirupati Laddu.
  - In Maharashtra- Mahabaleshwar strawberry.

Benefits

- Place names are sometimes used to identify a product. Well-known examples include "Champage", "Scotch", "Tequila", and "Roquefort" cheese, 'Basmati' rice and 'Darjeeling' Tea. Wine and spirits makers are particularly concerned about the use of place-names to identify products .
- In respect of all geographical indications, interested parties must have legal means to prevent use of indications, which mislead the public as to the geographical origin of the good, and use which constitutes an act of unfair competition.
- Protection for Wines and Spirits: Article 23 provides that interested parties must have the legal means to prevent the use of a

geographical indication identifying wines for wines not originating in the place indicated by the geographical indication. This applies even where the public is not being misled, there is no unfair competition and the true origin of the good is indicated or the geographical indication is accompanied by expressions, such as "kind", "type", "style", "imitation" or the like. Similar protection must be given to geographical indications identifying spirits, when used on spirits. Protection against registration of a trademark must be provided accordingly.

Ten organic and exotic agricultural products from North-East India were accorded geographical indication (GI) registration tag. GI tag will help to protect these exclusive special local crops and pave way for better branding and marketing of these products both, in domestic and international market. Union Government owned North Eastern Regional Agricultural Marketing Corporation Limited (NERAMAC) had played important role in getting GI registry. North Eastern Council (NEC) provided the financial support to this initiative.

Geographical Indication (GI) accorded products for North-East India are:

1. Assam Karbi Anglong Ginger.
2. Assam Tezpur Litchi.
3. Assam Joha Rice
4. Meghalaya Khasi Mandarin.
5. Sikkim Large Cardamom.
6. Mizoram Bird Eye Chilly.
7. Manipur Kachai Lemon.
8. Tripura Queen Pine apple.
9. Arunachal Orange.
10. Nagaland Tree Tomato

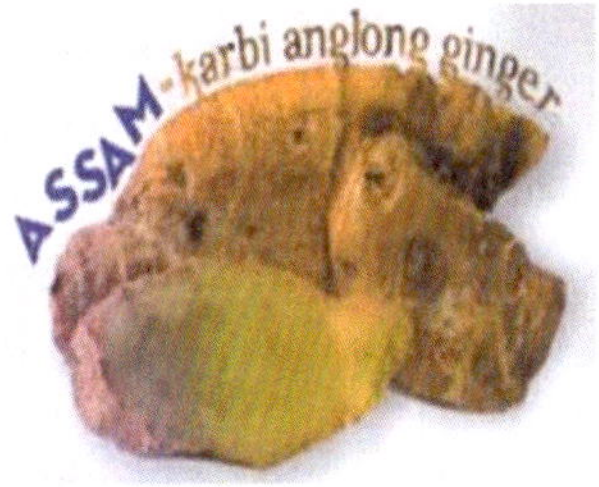
ASSAM-karbi anglong ginger

Tripura

Khasi Mandarin

Tezpur Litchi
Assam

joha
RICE OF ASSAM

Mizoram Bird Eye Chilly

ARUNACHAL PRADESH

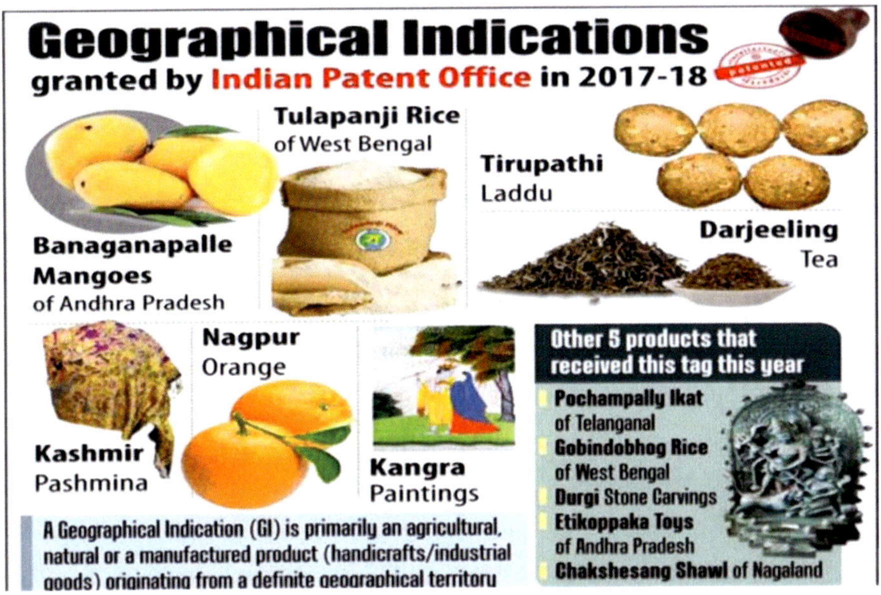
Geographical Indications
granted by Indian Patent Office in 2017-18
Tulapanji Rice
of West Bengal
Tirupathi
Laddu
Banaganapalle
Mangoes
of Andhra Pradesh
Darjeeling
Tea
Nagpur
Orange
Kashmir
Pashmina
Kangra
Paintings
Other 5 products that received this tag this year
Pochampally Ikat of Telanganal
Gobindobhog Rice of West Bengal
Durgi Stone Carvings
Etikoppaka Toys of Andhra Pradesh
Chakshesang Shawl of Nagaland
A Geographical Indication (GI) is primarily an agricultural, natural or a manufactured product (handicrafts/industrial goods) originating from a definite geographical territory

Laws relating to Geographical Indication of Goods

Geographical Indications of Goods (Registration and Protection) Act, 1999 and The Geographical Indications of Goods (Registration and Protection) Rules, 2002 deal with registration and better protection of geographical indications relating to goods. The primary purpose of this Act is to provide legal protection to Indian Geographical Indications, which in turn boost exports. Registration of Geographical indication promotes economic prosperity of producers of goods produced in a geographical territory.

Industrial designs

- An industrial design constitutes the ornamental or aesthetic aspect of an article.
- Owners of protected designs must be able to prevent the manufacture, sale or importation of articles bearing or embodying a design, which is a copy of the protected design.
- It consists of :
  - Three-dimensional features, such as the shape of a product;
  - Two-dimensional features, such as ornamentation, patterns, lines or colour of a product.
- Industrial designs are applied to a wide variety of industrial products and handicrafts: from technical and medical instruments to watches, jewellery and other luxury items; from house wares and electrical appliances to vehicles and architectural structures.
- To grant the owner of a protected industrial design the right to prevent third parties not having the owner's consent from making, selling or importing articles bearing or embodying a design, which is a copy, or substantially a copy, of the protected design, when such acts are undertaken for commercial purposes.

Trade Secrets

- Usually these are manufacturing or industrial secrets and commercial secrets.
- These include sales methods, distribution methods, consumer profiles, advertising strategies, lists of suppliers and clients, and

manufacturing processes. Contrary to patents, trade secrets are protected without registration.

Layout Design for Integrated Circuits

- The aim of the Semiconductor Integrated Circuits Layout-Design Act, 2000 is to provide protection of Intellectual Property Right (IPR).
- The main focus of SICLD Act is to provide for routes and mechanism for protection of IPR in Chip Layout Designs created and matters related to it.

Protection of New Plant Variety

Plant breeders' rights (PBR), also known as **plant variety rights** (PVR), are rights granted to the breeder of a new variety of plant that give the breeder exclusive control over the propagating material (including seed, cuttings, divisions, tissue culture) and harvested material (cut flowers, fruit, foliage) of a new variety for a number of years. With these rights, the breeder can choose to become the exclusive marketer of the variety, or to license the variety to others. In order to qualify for these exclusive rights, a variety must be new, distinct, uniform and stable. A variety is:

- New, if it has not been commercialized for more than one year in the country of protection;
- Distinct, if it differs from all other known varieties by one or more important botanical characteristics, such as height, maturity, colour, etc.;
- Uniform, if the plant characteristics are consistent from plant to plant within the variety;
- Stable, if the plant characteristics are genetically fixed and therefore, remain the same from generation to generation, or after a cycle of reproduction in the case of hybrid varieties.

The breeder must also give the variety an acceptable "denomination", which becomes its generic name and must be used by anyone who markets the variety. Typically, plant variety rights are granted by national offices, after examination. Seed is submitted to the plant variety office, which grow it for one or more seasons, to check that it is distinct, stable, and uniform.

Duration of protection of a registered plant variety

The duration of protection of registered varieties is different for different crops, which are as below:

1. For trees and vines - 18 years.
2. For other crops - 15 years.
3. For extant varieties - 15 years from the date of notification of that variety by the Central Government under section 5 of the Seeds Act, 1966.

Annual renewal fees are required to maintain the rights.

Breeders can bring suit to enforce their rights and can recover damages for infringement. Plant breeders' rights contain exemptions from infringement that are not recognized under patent law. Commonly, there is an exemption for farm-saved seed. Farmers may store the production in their own bins for their own use as seed, but this does not necessarily extend to brown-bag sales of seed. Further sales for propagation purposes are not allowed without the written approval of the breeder. There is also a breeders' exemption (research exemption in the 1991 Act) that allows breeders to use protected varieties as sources of initial variation to create new varieties of plants (1978 Act), or for other experimental purposes (1991 Act). There is also a provision for compulsory licensing to assure public access to protected varieties, if the national interest requires it and the breeder is unable to meet the demand.

There is tension over the relationship between patent rights and plant breeder's rights. There has been litigation in Australia, the United States, and Canada over the overlap between such rights. Each of these cases was decided on the principle that patents and plant breeders' rights were overlapping and not mutually exclusive. Thus, the exemptions from infringement of plant breeders' rights, such as the saved seed exemption, do not create corresponding exemptions from infringement of the patents covering the same plants. Likewise, acts that infringe the plant breeders' rights, such as exportation of the variety, would not necessarily infringe a patent on the variety, which only allows the patent owner to prohibit making, using or selling the patented invention.

Benefits of Registered Intellectual Property Rights

- Barrier to market entry- The owner can use the registered intellectual property right to create a barrier preventing or discouraging competition.
- Out-sourcing manufacture- If you plan to use out-sourced manufacturing to a company without first applying for registered intellectual property rights what will stop that company taking your idea and exploiting it themselves? A registered intellectual property right can prevent this.
- Cheap imports and counterfeit and pirated goods- If you have a registered intellectual property right, we can prevent importation of products that infringe your registered intellectual property rights.
- Provide protection against a competitor directly copying the idea.
- IP rights help business maintain their long-term competitive edge. Registered IP ensures that entrepreneurs get all the financial benefits from their ideas.
- Registered IP gives consumers confidence that products meet appropriate standards and quality.
- Registered IP is an asset. It helps convince financial institutions to invest in a business, enabling more money to be raised for development.

Positive Aspects of IP Rights

The primary, well-known function of an IP right is to give its holder a competitive advantage in its commercial activities, by preventing unauthorised exploitation by thirds. This is especially important for SMEs, which IP rights provide with powerful weapons to compete with much larger companies.

However, IP rights have other, less frequently mentioned benefits, including:

1. Providing guarantees regarding the quality & safety of products

Many counterfeit products place our children's and citizens' safety or health at risk, for instance, where vehicle spare parts or drugs are concerned. Enforcing IP rights in respect of such products guarantees at least that the products' origin is known and that the products are genuine, whereas counterfeit products often do not comply with the applicable safety standards. This is especially true for trademarks, but patent licensing contracts, for instance, may also include quality insurance clauses.

2. Enabling indirect exploitation

Where a company has protected its products (or processes, etc.) by IP rights, it can derive revenues not only from their direct exploitation (by that company), but also from their indirect exploitation by third parties, under licensing contracts. These additional indirect revenues sometimes exceed the profits resulting from the direct exploitation, especially as they do not require additional internal manufacturing capacities. Such an approach may, therefore, be particularly relevant for SMEs. It is also important for universities and public research centres, which usually do not have any direct exploitation activities.

3. Cost-free mechanisms

While certain procedures required for the registration of IP rights are considered to be expensive, in particular by SMEs, it should be noted that certain IP rights can be enjoyed without any formal procedure and without paying any official fees. This is in particular the case for copyright and for unregistered designs.

4. Dissemination of technical information

Even where a company (or university, etc.) does not intend to protect its own inventions, its staff (researchers, etc.) can still make use of patent information.

Patents are the most prolific and up-to-date source of technological information, and contain detailed technical information, which often cannot be found anywhere else: it is estimated that upto 80% of current technical knowledge can only be found in patent documents. Moreover, this information is rapidly available, as most patent applications are published 18 months after the first filing.

There are good reasons to search patent literature:

- Avoid duplication of R&D efforts and spending (it is estimated that upto 30% of all expenditure in R&D is wasted on re-developing existing inventions).

- Find solutions to technical problems (especially as the majority of all patents
- Around 85% – are no longer in force, a vast number of inventions is, thus, available for free).
- Gather business intelligence (as patents not only reveal technological information areas, but also make it possible, at a very early stage, to identify potential competitors, customers and partners; to monitor the innovation strategies of competitors; etc.).

5. Facilitating technology transfer

Patents often constitute a convenient means to not only protect but also describe in a very accurate way technologies, which are the subject of technology transfer and similar agreements (licensing, assignment, etc.). This "technology packaging"/trade facilitation function justifies that patents have sometimes been considered as the "currency" of the knowledge-based economy. (To some extent, the same reasoning also applies to IP rights other than patents).

6. "Open source" relies on IPR

Open source mechanisms are becoming popular in certain sectors, such as software (cf. GPL licences, etc.). While the common perception is that such mechanisms are characterized by the absence of any IP protection, it is worth noting that a typical GPL (General Public) licence actually relies on IP rights, as it is typically a copyright license which remains valid as long as certain conditions are complied with (e.g. freedoms received by the licensee must be passed on to subsequent users, even where the software is modified).

7. Collateral to obtain financing

As intangible assets, IP rights often play an instrumental role for SMEs (including start-ups and spin-offs) trying to convince third parties to provide financing to them (equity investment, loan granting, etc.).

IPRs provide a basis for businesses to:

1. Prevent others from copying their products or using their innovations
2. Gain revenues through franchising, licensing or other IP transactions

3. Create a strong brand identity- by product differentiation using one or more types of IPRs
4. Obtain valuable competitive intelligence(information)- analyzing competitors
5. Obtain financing or venture capital -IP assets may be used as collateral from banks
6. Increase their commercial value
7. Enter new markets
8. Engage in different types of collaborative business partnerships.
9. Reducing the risk of IPR infringement; by owning or licensing an IPR the businesses can operate freely and the risk of infringing IPRs using others technologies, trademarks or designs is reduced.

Overview of laws related to intellectual property rights in India

The Rules and Laws governing Intellectual Property Rights in India are as follows:

1. The Copyright Act, 1957, The Copyright Rules, 1958 and International Copyright Order, 1999
2. The Patents Act, 1970 The Patents Rules, 2003, The Intellectual Property Appellate Board (Patents Procedure) Rules, 2010 and The Patents (Appeals and Applications to the Intellectual Property Appellate Board) Rules, 2011
3. The Trade Marks Act, 1999, The Trade Marks Rules, 2002, The Trade Marks (Applications and Appeals to the Intellectual Property Appellate Board) Rules, 2003 and The Intellectual Property Appellate Board (Procedure) Rules, 2003
4. The Geographical Indications of Goods (Registration and Protection) Act, 1999 and The Geographical Indications of Goods (Registration and Protection) Rules, 2002
5. The Designs Act, 2000 and The Designs Rules, 2001
6. The Semi-conductors Integrated Circuits Layout-Design Act, 2000 and The Semi-conductors Integrated Circuits Layout-Design Rules, 2001

7. The Protection of Plant varieties and Farmers' Rights Act, 2001 and The Protection of Plant varieties and Farmers Rights' Rules, 2003
8. The Biological Diversity Act, 2002 and The Biological Diversity Rules, 2004
9. Intellectual Property Rights (Imported Goods) Rules, 2007.

Conclusions

The issue of respecting of intellectual property rights has many dimensions. It is a moral issue of respecting human rights, respecting his/her creative works. It is also an issue related to progress and development, which stands at the center of the social and economic development of a country. It is related with fair competition, the engine of economic development of any nation. It is also an obligation for countries, like Albania for the path towards European Union, not only formally fulfilling legal obligations in harmonizing the legislation but also requiring the enforcement of the law. It is also an issue that goes beyond institutions, and law, it involves the civil society, the consumers, it is a matter of public awareness.

❑❑❑

Chapter **4**

Indian Legislations for the Protection of Various Types of Intellectual Properties

Intellectual Property System in India As discussed above, historically the first system of protection of intellectual property came in the form of (Venetian Ordinance) in 1485. In India, Patent Act was introduced in the year 1856, which remained in force for over 50 years, which was subsequently modified and amended and was called. "The Indian Patents and Designs Act, 1911". After Independence, a comprehensive bill on patent rights was enacted in the year 1970 and was called "The Patents Act, 1970". In India:

- Copyrights were regulated under the Copyright Act, 1957;
- Patents under Patents Act, 1970;
- Trade marks under Trade and Merchandise Marks Act, 1958; and
- Designs under Designs Act, 1911.

With the establishment of WTO and India being signatory to the Agreement on Trade-Related Aspects of Intellectual Property Rights (TRIPS), several new legislations were passed for the protection of intellectual property rights to meet the international obligations. These included:

- Trade Marks, called the Trade Mark Act, 1999;
- Designs Act, 1911 was replaced by the Designs Act, 2000;
- The Copyright Act, 1957 amended a number of times, the latest is called Copyright (Amendment) Act, 2012;
- Latest amendments made to the Patents Act, 1970 in 2005.

Besides, new legislations on geographical indications and plant varieties were also enacted. These are called:

- Geographical Indications of Goods (Registration and Protection) Act, 1999
- Protection of Plant Varieties and Farmers' Rights Act, 2001.

Deveopment of TRIPS Complied Regime in India

The establishment of WTO as a result of institutionalization of international framework of trade calls for harmonization of several aspects of Indian Law relating to Intellectual Property Rights.

Patents Act, 1970

- For the purpose of reduction of distortions and impediments to international trade and promotion of effective and adequate protection of intellectual property rights.
- The Patents Act, 1970 has been amended in the year 1995, 1999, 2002 and 2005 to meet its obligations under the TRIPS agreement.
- The amendments were also aimed at making the Act a modern, harmonized and user-friendly legislation to adequately protect national and public interests, while simultaneously meeting India's international obligations under the TRIPS Agreement.
- The Patent Amendment Act, 2005 is now fully in force and operative.

Trade Mark Act, 1999

- The law of trademarks is also now modernized under the Trademarks Act of 1999. A trademark is a special symbol for distinguishing the goods offered for sale or otherwise put on the market by one trader from those of another.
- In India, the trademarks have been protected for over four decades as per the provisions of the Trade and Merchandise Mark (TMM) Act of 1958.
- In December, 1998 India acceded to the Paris Convention. Meanwhile, the modernisation of Trade and Merchandise Marks Act, 1958 had been taken up keeping in view the current developments in trading and commercial practices, increasing globalisation of trade and industry, the need to encourage investment flows and transfer of technology, need for

simplification of trademark management system and to give effect to important judicial decisions.

- To achieve these purposes, the Trademarks Bill was introduced in 1994.
- A comprehensive review was made of the existing laws in view of the developments in trading and commercial practices, and increasing globalization of trade and industry.
- The Trademarks Bill of 1999 was passed by Parliament that received the assent of the President on 30th December, 1999, as Trade Marks Act, 1999 thereby replacing the Trade and Merchandise Mark Act of 1958.

  The important salient features of the Act inter-alia include:

  - It broadens the definition of infringement of a registered trademark to include action against the unauthorized use of a confusingly similar mark, not only in respect of the goods and services covered by registration, as was previously the case, but also in respect of goods and services, which are so similar that a likelihood of deception or confusion exists.
  - An action for infringement will also be available against the unauthorised use of a mark in relation to dissimilar goods, if such mark is similar to a registered mark that is well known in India and the interest of the owner is likely to be affected adversely. The remedy for infringement of a trademark is also strengthened under the new law by authorising the police with the power to seize infringing articles without a warrant.

The Designs Act, 2000

- The Designs Act of 1911 has been replaced by the Designs Act, 2000.
- A need was felt to provide more efficient legal system for the protection of industrial designs in order to ensure effective protection to registered designs, and to encourage design activity to promote the design element in an article of production.
- In this backdrop, the Designs Act, 2000 has been enacted essentially to balance these interests and to ensure that the law does not unnecessarily extend, protection beyond what is

necessary to create the required incentive for design activity, while removing impediments to the free use of available designs.

- Today, industrial design has become an integral part of consumer culture, where rival articles compete for consumer's attention.
- The salient features of the Design Act, 2000 are as under:
  - Enlarging the scope of definition of the terms "article", "design" and introduction of definition of "original".
  - Amplifying the scope of "prior publication".
  - Introduction of provision for delegation of powers of the Controller to other officers and stipulating statutory duties of examiners.
  - Provision of identification of non-registrable designs.
  - Provision for substitution of applicant before registration of a design.
  - Substitution of Indian classification by internationally followed system of classification.
  - Provision for inclusion of a register to be maintained on computer as a Register of Designs.
  - Provision for restoration of lapsed designs.
  - Provisions for appeal against orders of the Controller before the High Court instead of Central Government.
  - Revoking of period of secrecy of two years of a registered design.
  - Providing for compulsory registration of any document for transfer of right in the registered design.
  - Introduction of additional grounds in cancellation proceedings and provision for initiating the cancellation proceedings before the Controller in place of High Court. (m) Enhancement of quantum of penalty imposed for infringement of a registered design.
  - Provision for grounds of cancellation to be taken as defence in the infringement proceedings to be in any court not below the Court of District Judge.
  - Enhancing initial period of registration from 5 to 10 years, to be followed by a further extension of five years.

- Provision for allowance of priority to other convention countries and countries belonging to the group of countries or inter-governmental organizations apart from United Kingdom and other Commonwealth Countries.
- Provision for avoidance of certain restrictive conditions for the control of anticompetitive practices in contractual licenses.

The Geographical Indications of Goods (Registration and Protection) Act, 1999

- Indian geographical indications had been misused by persons outside India to indicate goods not originating from the named locality in India. Patenting turmeric, neem and basmati are the instances which drew a lot of attention towards this aspect of the Intellectual property.
- Mention should be made that under the Agreement on Trade Related Aspects of Intellectual Property Rights (TRIPS), there is no obligation for other countries to extend reciprocal protection unless a geographical indication is protected in the country of its origin.
- India did not have such a specific law governing geographical indications of goods, which could adequately protect the interest of producers of such goods.
- To cover up such situations it became necessary to have a comprehensive legislation for registration and for providing adequate protection to geographical indications and accordingly the Parliament has passed a legislation, namely, the Geographical indication of Goods (Registration and Protection) Act, 1999.
- The legislation is administered through the Geographical Indication Registry under the overall charge of the Controller General of Patents, Designs and Trade Marks. The salient features of this legislation are as under:
    - Provision of definition of several important terms, like "geographical indication", "goods", "producers", "packages", "registered proprietor", "authorized user", etc.
    - Provision for the maintenance of a Register of Geographical Indications in two parts-Part A and Part B and use of computers etc. for maintenance of such Register. While Part

A will contain all registered geographical indications, Part B will contain particulars of registered authorized users.

- Registration of geographical indications of goods in specified classes.
- Prohibition of registration of certain geographical indications.
- Provisions for framing of rules by Central Government for filing of application, its contents and matters relating to substantive examination of geographical indication applications.
- Compulsory advertisement of all accepted geographical indication applications and for inviting objections.
- Registration of authorized users of registered geographical indications and providing provisions for taking infringement action either by a registered proprietor or an authorized user.
- Provisions for higher level of protection for notified goods.
- Prohibition of assignment, etc. of a geographical indication as it is public property.
- Prohibition of registration of geographical indication, as a trademark.
- Appeal against Registrar's decision would be to the Intellectual Property Board established under the Trade Mark legislation.
- Provision relating to offences and penalties.
- Provision detailing the effects of registration and the rights conferred by registration.
- Provision for reciprocity powers of the registrar, maintenance of Index, protection of homonymous geographical indications, etc.

Copyright Act, 1957

- Copyright in India is governed by Copyright Act, 1957.
- This Act has been amended several times to keep pace with the changing times.
- As per this Act, copyright grants author's lifetime coverage plus 60 years after death.

- Copyright and related rights on cultural goods, products and services, arise from individual or collective creativity.
- All original intellectual creations expressed in a reproducible form will be connected as "works eligible for copyright protections".
- Copyright laws distinguish between different classes of works, such as literary, artistic, musical works and sound recordings and cinematograph films.
- The work is protected irrespective of the quality thereof and also when it may have very little in common with accepted forms of literature or art.
- Copyright protection also includes novel rights which involve the right to claim authorship of a work, and the right to oppose changes to it that could harm the creator's reputation.
- The creator or the owner of the copyright in a work, can enforce his right administratively and in the courts by inspection of premises for evidence of production or possession of illegally made "pirated" goods related to protected works.
- The owner may obtain court orders to stop such activities, as well as seek damages for loss of financial rewards and recognition.
- A vital field which gets copyright protection is the computer industry.
- The Copyright Act, 1957 was amended in 1984 and computer programming was included with the definition of "literary work.'
- The new definition of "computer programme" introduced in 1994, means a set of instructions expressed in works, codes or in any other form, including a machine readable medium, capable of causing a computer to perform a particular task or achieve a particular result.
- The greatest fear and challenges to the copyright industry is the piracy of works whether, books, musical works, films, television programmes or computer software or computer database.
- The special nature of infringement of copyrights in computer programmes has again been taken note of by the Copyright (Amendment) Act, 1994 by inserting a new section 63 B.

- The new section provides that any person who knowingly makes use on a computer of an infringed copy of a computer programme will be punishable with imprisonment for a term of not less than seven days, which may extend to three years and with a fine of not less than Rs. 50,000/- and which may extend to Rs. 2,00,000/-. Proviso to section 63 B, however, provides that where computer programme has not been used for gain or in the course of trade or business, the court may at its discretion and for reasons mentioned in the judgement not impose any sentence of imprisonment and impose only fine upto Rs. 50,000/-.
- The Copyright (Amendment) Act, 1999 makes it free for purchaser of a gadget/equipment to sell it onwards, if the item being transacted is not the main item covered under the Copyright Act.
- With this amendment India has updated the Act to meet the concerns of the copyright industries mainly consisting of Book Industry, Music Industry, Film and Television Industry, Computer Industry and Database Industry.

The Protection of Plant Varieties and Farmers' Rights Act (PPVFR Act), 2001

- The issue of enacting a law relating to Plant Varieties Protection and Farmers' Rights in India assumed importance particularly in the wake of TRIPS agreement under WTO, which seeks to promote effective protection of Intellectual Property Rights in all fields of technology.
- Article 27 of TRIPS Agreement defines patentable subject-matter and requires member countries to provide for the protection of plant varieties whether by patenting or by an effective sui generis system or by any combination thereof.
- With a view to provide for the establishment of an Authority to give an effective system of protection of the rights of plant breeders and farmers, and to encourage the development of new varieties of plants and to give effect to the provisions of TRIPS Agreement, the Government enacted the Protection of Plant Varieties and Farmers' Right Act, 2001.
- This Act seeks to stimulate investment for research and development both in the public and private sectors for the

development of new plant varieties by ensuring appropriate returns on such investments.

- It also seeks to facilitate the growth of the seed industry in the country through domestic and foreign investment to ensure the availability of high quality seeds and planting material to Indian farmers.
- It also recognizes the role of farmers as cultivators and conservers and the contribution of traditional, rural and tribal communities to the country's agro biodiversity by rewarding them for their contribution through benefit sharing and protecting the traditional rights of the farmers.
- The Act also provides for setting up of the Protection of Plant Varieties and Farmer's Rights Authority to promote and develop new varieties of plants and promotes rights of the farmers and breeders.

The Semi-Conductor Integrated Circuits Layout Design Act, 2000

- Electronics and Information technology is one of the fastest growing sectors that has played a significant role in world economy.
- This is primarily due to the advancements in the field of electronics, computers and telecommunication.
- Microelectronics, which primarily refers to Integrated Circuits (ICs) ranging from, Small Scale Integration (SSI) to Very Large Scale Integration (VLSI) on a semiconductor chip - has rightly been recognized as a core, strategic technology world-over, especially for Information Technology (IT) based society.
- Design of integrated circuits requires considerable expertise and effort depending on the complexity.
- Therefore, protection of Intellectual Property Rights (IPR) embedded in the layout designs is of utmost importance to encourage continued investments in R & D to result in technological advancements in the field of microelectronics.
- The practice of providing protection through the methods of Copyright, Patents did not appropriately accommodate the requirements of Intellectual Property Rights protection for the Layout-Designs of Integrated Circuits.

- This was because of the fact that in the context of Layout Designs, the concept of "originality" is of utmost significance, whether it is a "novelty or not". While the Patent Law requires that the idea should be original as well as novel, the copyright law is too general to accommodate the original ideas of scientific creation of Layout-Designs of Integrated Circuits.
- In view of the above, the necessity for providing protection for Layout-Designs of Integrated Circuits was felt to reward and encourage an adequate level of investment of human, financial and technological resources.
- The Majority of countries that attach significance to protection of Intellectual Property Rights in the Semi-conductor Integrated Circuits provides for sui generis system of protection of Layout-Designs of Integrated Circuits, which is usually contained in a separate Act.
- Trade Related Intellectual Property Rights (TRIPS) Agreement under WTO contains provisions with regard to setting up of standards concerning availability, scope and use of Intellectual Property Rights, Geographical Indications, Layout-Design of Integrated Circuits, etc.
- Therefore, the Government enacted the Semi-conductor Integrated Circuit LayoutDesigns Act, 2000 providing for protection of Semiconductor Integrated Circuits Layout-Designs by process of registration, mechanism for distinguishing Layout-Designs which can be protected, rules to prohibit registration of Layout-Designs, which are not original and/or which have been commercially exploited, period for protection, provisions with regard to infringement, payment of royalty for registered Layout-Design, provisions for dealing with willful infringement by way of punishment, appointing a Registrar for registering the Layout Designs and mechanism of Appellate Board.
- To conclude, the various modifications and amendments to earlier Intellectual Property Laws are an indication of India's move towards new IPR regime so as to prepare ourselves for the global trade competition.
- Usually these are manufacturing or industrial secrets and commercial secrets. These include sales methods, distribution methods, consumer profiles, and advertising strategies, lists of suppliers and clients, and manufacturing processes.

- A trade secret can be protected for an unlimited period of time but a substantial element of secrecy must exist so that except by the use of improper means, there would be difficulty in acquiring the information. Considering the vast availability of traditional knowledge in the country, the protection under this will be very crucial in reaping benefits from such type of knowledge.

Utility Models

- A utility model is an exclusive right granted for an invention, which allows the right holder to prevent others from commercially using the protected invention, without his authorization for a limited period of time.
- In its basic definition, which may vary from one country (where such protection is available) to another, a utility model is similar to a patent.
- In fact, utility models are sometimes referred to as "petty patents" or "innovation patents."
- Only a small but significant number of countries and regions provide the option of utility model protection.
- At present, India does not have legislation on Utility models.
- The main differences between utility models and patents are the following:
    - The requirements for acquiring a utility model are less stringent than for patents.
    - While the requirement of "novelty" is always to be met, that of "inventive step" or "non-obviousness" may be much lower or absent altogether.
    - In practice, protection for utility models is often sought for innovations of a rather incremental character, which may not meet the patentability criteria. The term of protection for utility models is shorter than for patents and varies from country to country (usually between 7 and 10 years without the possibility of extension or renewal).
    - In most countries, where utility model protection is available, patent offices do not examine applications as to substance prior to registration.

- This means that the registration process is often significantly simpler and faster, taking on an average six months.
- Utility models are much cheaper to obtain and to maintain.
- In some countries, utility model protection can only be obtained for certain fields of technology, and only for products but not for processes.
- Utility models are considered suitable particularly for SMEs that make "minor" improvements to, and adaptations of, existing products.
- Utility models are primarily used for mechanical innovations.

□□□

Chapter 5

Convention on Biological Diversity, International Treaty on Plant and Animal Genetic Resources for Food and Agriculture

The International Treaty on Plant Genetic Resources for Food and Agriculture, referred to as the Plant Treaty, was approved on 3 November, 2001 by Members of the Food and Agriculture Organization (FAO), headquartered in Rome, Italy. The FAO is an agency of the United Nations, headquartered in New York City, New York. The Plant Treaty established international standards for the conservation and exchange of plant genetic material between participating countries. Plant genetic material is a term for plant germplasm, the physical material used by plants to reproduce them, and the term connotes seeds, vegetative propagations, and DNA. Plant genetic resources are the collective genetic diversity of plant species in the laboratory, farm, and field. They are described as resources because of their value for food and agricultural purposes.

The Plant Treaty aimed to promote the conservation and sustainable use of plant genetic resources for global food security. Conservation includes both on-farm conservation and seed banks. A result of prolonged discussion within the FAO, the Plant Treaty established a principle of Farmers' Rights, which recognized farmers as holders of traditional knowledge and stewards of agricultural biodiversity. Next, the Plant Treaty established a multilateral system for open access to plant genetic materials. The open access policy is meant to encourage countries to establish laws on plant genetic materials that would assure the rights of both, farmers and plant breeders.

The Plant Treaty established standards for fair and equitable access to plant genetic materials and benefits sharing between contracting parties. For access to plant genetic materials, the Plant Treaty established that seed banks and other international seed collections must provide facilitated access to the banks, when a contracting party requests access. However, this provision does not apply to private organizations, such as seed companies and pharmaceutical companies. The Plant Treaty covers the open exchange of materials that are exclusively for food use. According to the treaty, parties that commercialize and profit from products of plant genetic materials should share their profits by giving a percentage of the profits to a common fund that provides financial and technical support for farmers and local communities, especially in developing nations.

The FAO's International Undertaking on Plant Genetic Resources (1983) preceded the Plant Treaty. A voluntary agreement between FAO member states, the International Undertaking established a Commission on Plant Genetic Resources to oversee international collection and storage of plant genetic material. It aimed to protect the legal rights of plant breeders and didn't address farmers' rights. Scholars note that the International Undertaking did not have much impact on national policies because of its lack of legal force, and that it was ultimately ineffective. The International Undertaking aimed to improve international access to plant genetic materials, but it didn't foster a consensus about ensuring both, plant breeders and farmers' rights. Later revisions to the International Undertaking in 1989 and 1991, attempted to address these concerns; however, the 1992 Convention for Biological Diversity overshadowed the International Undertaking. The International Undertaking still provided a framework for international oversight of plant genetic resources, on which the FAO began its negotiations for a new treaty. The International Undertaking formally existed until the Plant Treaty in 2001.

The Convention on Biological Diversity (CBD), signed by 193 countries in 1992 and enacted in 1993, was the first legally binding international treaty to address the international exchange of plant and animal genetic resources. The CBD established a framework for the conservation and sustainable use of biodiversity. It also created a system of access and benefit sharing for the global trade of plant and animal genetic resources. While the Plant Treaty is independent of the CBD, the Plant Treaty reflects the CBD's principles of conservation, sustainable use of biodiversity, access to resources, and benefit sharing.

The CBD didn't address several issues of plant genetic resources for food and agriculture. First, the CBD gave special rights to the country of origin of the genetic material. In the case of agricultural crops, there is often no single country of origin as a result of thousands of years of breeding between plants. The CBD complicated the legal status of global seedbanks by asserting that the laws of individual nations governed the ownership of plant genetic resources. After years of debate, the world's largest and most diverse collection of plant genetic resources, held by the eleven Consultative Group for International Agriculture Research (CGIAR) Centers, was brought under FAO policy, which prohibited nations from claiming ownership of plant genetic resources, and which established CGIAR Centres as trustees of the collections. Additionally, the CBD did not address development and management of improved plant genetic material, or germplasm that is augmented through plant breeding, biotechnology, and other scientific techniques. Some consider improved plant genetic material as man-made biodiversity, as opposed to raw plant genetic material. After the CBD was passed, in 1993, the FAO set out to harmonize the International Undertaking with the gaps left by the CBD. An intermediate step was the creation of the Global Plan of Action for Conservation and Sustainable Use of Plant Genetic Resources for Food and Agriculture in 1996, which was voluntarily adopted by 150 countries, and later contributed to language of the Plant Treaty.

Based on the precedents set by the FAO's International Undertaking and the Convention on Biological Diversity, the Plant Treaty established standards for the international exchange of plant genetic materials for food and agricultural uses. Unlike the CBD, which mandates exchange agreements between individual countries, the Plant Treaty is multilateral, meaning that agreements are multinational, where one decision applies to all participating countries equally. This multilateral agreement reconciles the global nature of plant genetic resources that the CBD lacked. For example, under the CBD, if the US wanted to develop a new drug based on a plant from Brazil, the US would first negotiate an agreement with Brazil for sharing benefits based on the sales of the final product. Under the Plant Treaty, if the US wanted to obtain plant genetic material from a seedbank in Brazil for an agricultural purpose, it would enter into an international agreement for use of that material, rather than a contract with Brazil. The multilateral agreement also addresses the difference between raw and improved plant genetic material, including domesticated crops that were difficult to cover under the CBD.

Negotiations on the Plant Treaty took place over seven years, beginning in 1993, as mandated at an FAO conference. While the main negotiators at the conference were the 192 member states of the FAO, the private sector, such as seed companies, also participated. Activist groups and non-governmental organizations were also involved, and some commentators note that the FAO was partial to the activist perspective. These negotiations included not just formal disagreements over the Plant Treaty, but long-standing conflicts over the intellectual property rights of plant genetic resources that began in the late 1970s. The Plant Treaty overcomes at least twenty-two years of disagreement between different organizations, countries, activists, and interest groups. As an agency within the United Nations, FAO had long pushed for the creation of an international network centralized within the FAO.

Chaired by scientist and FAO council member Monkombu Sambasivan Swaminathan, many of the negotiations over the Plant Treaty occurred between 1997 and 2001. An informal meeting of experts in Montreux, Switzerland, in 1999 between the chairman and his supporters influenced the final negotiations. Forty delegations participated in the FAO negotiations, while the US was conspicuously absent. Observers note that developed countries, such as those within the European Union, dominated the negotiations, and that most of the conflicts were between developed and developing nations, especially over intellectual property rights. Representatives from developing nations, such as India, argued that the Treaty negotiations could result in exploitation of plant genetic resources by developed nations. Consequently, in an attempt to accommodate all parties, some of the Plant Treaty's language is ambiguous about intellectual property rights.

Once ratified by forty member countries, the required number for enacting treaties, the Plant Treaty went into effect on 29 June, 2004. As of 2012, 126 countries have signed and ratified the treaty, meaning that they are Contracting Parties of the Plant Treaty. To enforce the Treaty, The Governing Body was created in 2004. The Secretary and Governing Body of the Plant Treaty are located within the FAO, headquartered in Rome, Italy. The FAO's Governing Body of the Plant Treaty has met about every two years. These meetings have addressed implementation of and compliance with the Plant Treaty, relationships with other international organizations, and the status of funds acquired through benefit sharing.

Some scholars have criticized what they claim is a somewhat limited list of crops actually covered by the Plant Treaty. Known as Annex I crops, thirty-five varieties of crops and crop families are included in the Plant Treaty. This list does not cover some agriculturally important crops, for example, soyabeans, as China engaged in political conflicts with the US at the time of treaty negotiation and refused to include soybeans on the list. Many other countries withheld specific crops from the Annex as a bargaining tool, because the Plant Treaty negotiations required consensus among the member nations for every plant. Later negotiations have specified protocols for the exchange of non-Annex I crops from international seed banks.

The Plant Treaty created and enforced the Standard Material Transfer Agreement (SMTA), a mandatory, legally binding agreement between parties exchanging plant genetic materials, for example, when a plant breeder wants to access a specific seed variety that is stored at a seedbank. The initial treaty did not specify the SMTA, but left this specification to later meetings of the Plant Treaty's Governing Body. The first meeting of the Governing Body in Madrid, Spain, in 12 through 16 June, 2006 established this SMTA. This legal agreement uses a mandatory template to outline standards of equal access for both, the providers and recipients of plant genetic material. The SMTA does not address other aspects of intellectual property rights, such as whether exchanged plant genetic materials can be patented. Intellectual property rights vary from country to country, although the World Trade Organization's Trade-Related Aspects of Intellectual Property Rights in 1994 established a system of international property rights for plant varieties. Despite some resistance from private seed companies that disagree with the SMTA, thousands of plant genetic material transactions have occurred using the SMTA, and only a small handful of private parties have refused to abide by the SMTA.

The 2006 meeting of the Plant Treaty's Governing Body also formed an official relationship with the Global Crop Diversity Trust. The Global Crop Diversity Trust, headquartered in Bonn, Germany, contributes to the funding strategy of the Plant Treaty. Cary Fowler, who helped negotiate the Plant Treaty for the Consultative Group for International Agricultural Research (CGIAR), is the Special Advisor and former

Executive Director of the Global Crop Diversity Trust. The Global Crop Diversity Trust helps administrate the Svalbarg Global Seed Vault, located in Norway. The Svalbard Global Seed Vault, opened in 2008, is a long-term storage facility for plant germplasm, and has gained international recognition.

By 2013, some have described some of the impacts of the Plant Treaty on the international exchange of plant genetic resources. First, by establishing an international standard, the Plant Treaty clarified much of the legal uncertainty that may have caused a decreased exchange of plant genetic materials in the 1990s. Between 1992 and 2002, the exchange of materials had declined as some countries withheld their genetic resources from international exchange. Since the Treaty's ratification in 1993, the number of global exchanges of plant genetic material has increased. Furthermore, the Plant Treaty established the SMTA that facilitates the transfer of plant genetic resources from the CGIAR's seed banks, especially for crop research and development. Some scholars have said that the Plant Treaty's multilateral system is more effective for supporting access and exchange of genetic materials than the Convention on Biological Diversity. Others claim that benefits are still limited, as private corporations may take advantage of these resources. At the national level, there are still barriers to implementing the Plant Treaty. Some countries lack infrastructure or political impetus to set national policies around plant genetic resources.

❑❑❑

Chapter 6

IPR & Bilodiversity

In simple terms, the diversity among various life forms within the Biosphere refers to biodiversity. Biodiversity is the foundation of life on Earth. It is crucial for the functioning of ecosystems, which provide us with products and services without which we cannot live. By changing biodiversity, we strongly affect human well-being and the well-being of every other living creature. Biodiversity is normally classified under 3 major categories: Lesson 1 Introduction 11 ecosystem diversity, representing the principal bio geographic regions and habitats; species diversity, representing variability at the level of families, genera and species; and genetic diversity, representing the large amount of variability occurring within a species. Diverse activities and actions have been taken by several stakeholders at local, state, national and international level to conserve/protect the valuable resource, such as biodiversity to draw the benefits accrued in it for the society. It is a well-established fact that developing countries are rich in the world's flora and fauna and 80 per cent of the earth's terrestrial biodiversity is confined to these countries, which is the "raw material" for biotechnology, i.e., genes, folk varieties, land races to develop new varieties by biotechnology. Until the advent of molecular biology and genetic engineering, the success of plant breeding depended on access to genetic variability within a species. Genetic engineering has, however, rendered the transfer of genes across sexual barriers possible and has, thus, enhanced the economic value of biodiversity. The developed countries are not rich in biogenetic resources but are better equipped in research and development. They use the biogenetic resources accessed from the developing countries. As a result, there is a beginning in the unprotected flow of genetic information from the developing countries to the capital-rich west, and a protected flow in the reverse direction mainly through patents and Plant Breeders' Rights (PBR). It has both, visible and invisible impacts. Genetic erosion is one of the most important invisible impacts that is in the long run manifested visibly with the loss of biodiversity. The Convention on Biological Diversity (CBD) 1992: Opened for signature at the Earth Summit in Rio de Janeiro in 1992, and entering into force in December 1993, the Convention on Biological Diversity is an international treaty for the conservation of biodiversity, the sustainable use of the components of biodiversity and the equitable sharing of the benefits derived from the use of genetic resources. The interface between biodiversity and intellectual property is shaped at the international level by several treaties and process, including at the WIPO, and the TRIPS Council of the WTO. With 193 Parties, the Convention has near universal participation among countries. The Convention seeks to address all threats to biodiversity and ecosystem services, including threats from climate change, through scientific assessments, the development of tools, incentives and processes, the transfer of technologies and good practices and the full and active involvement of relevant stakeholders including indigenous and local communities, youth, NGOs, women and the business community. The Cartagena Protocol on Biosafety is a subsidiary agreement to the Convention. It seeks to protect biological diversity from the potential risks posed by living modified organisms resulting from modern biotechnology. The treaty defines biodiversity as "the variability among living organisms from all sources including inter alia, terrestrial, marine and other aquatic ecosystems and the ecological complexes of which they are part; this includes diversity within species, between species and of ecosystems." The Convention re-affirms the principle of state sovereignty, which grants states sovereign rights to exploit their resources pursuant to their own environmental policies together with the responsibility to ensure that activities within their own jurisdiction or control do not cause damage to the environment of other states. The Biodiversity Convention also provides a general legal framework regulating access to biological resources and the sharing of benefits arising from their use. India is a party to the Convention on Biological Diversity (1992). The Convention on Biological Diversity establishes important principles regarding the protection of biodiversity, while recognizing the vast commercial value of the planet's store of germplasm. However, the expansion of international trade agreements establishing a global regime of intellectual property rights creates incentives that may destroy biodiversity, while undercutting social and economic development opportunities as well as cultural diversity. The member countries were pressurized to change their IPR laws to conform with the TRIPS agreement. 12 PP-IPRL&P India also followed the suit by placing in place legal frameworks for the management of biodiversity and Intellectual property laws. Following India's ratification of the Convention on Biological Diversity (CBD) at international level, the Biological Diversity Act, 2002 was adopted.

Biodiversity Act, 2002

India enacted The Biological Diversity Act, 2002 and The Biological Diversity Rules, 2004 to fulfill its commitments in the conservation of biological resources and associated knowledge as well as facilitating access to them in a sustainable manner and through a just process.

Various states have also enacted state specific Biological diversity rules, namely :

i) Kerala Biological Diversity Rules, 2005

ii) Sikkim State Biological Diversity Rules, 2006

iii) Nagaland Biological Diversity Rules, 2010

iv) Rajasthan Biological Diversity Rules, 2010

v) A.P. State Biological Diversity Rules, 2009

vi) West Bengal Biological Diversity Rules, 2005; U.P.

vii) State Biodiversity Rules, 2010

viii) Maharashtra Biological Diversity Rules, 2008, etc.

The Biodiversity Act, 2002 primarily addresses access to genetic resources and associated knowledge by foreign individuals, institutions or companies, to ensure equitable sharing of benefits arising out of the use of these resources and knowledge to the country and the people.

The Act has specific provisions about ownership of intellectual property rights associated with exploitation of biodiversity. Industries have to obtain prior consent of the National Biodiversity Authority before exploring the biodiversity in India. In the event of R&D based on exploitation of biodiversity and associated local knowledge, there is a provision for sharing of benefits of such work with the local community.

The Biological diversity Act of 2002 contains 65 sections under 12 chapters, while the Biological Diversity Rules of 2004 consists of 24 rules and one schedule.

❑❑❑

Chapter 7

Advantages and Disadvantages of Intellectual Property Rights

The concept of intellectual property is not an entirely new one, but it has become increasingly more important, as the economy worldwide has begun to shift into, what is known as an information economy. This shift has started to change what is considered valuable in private exchange. Instead of physical objects, or land, the most important form of property in this new world is intellectual, the creative products of the human mind.

Trademark

A sign that can distinguish the goods and services from one trader from those of another. Might include words, logos, pictures or a combination of these.

Advantage

Legally prevents others taking advantage of customer good will generated by your business by trading under your name (or something very similar).

Disadvantages

Having a registered trademark does not entitle you to the related internet domain names. The degree to which similar (not identical) trademarks infringe upon your business is a matter for debate in civil action with related cost.

Copyrights

Comes into effect immediately something that can be protected is created and 'fixed' insomeway e.g. on paper, on film, via sound recording, as an electronic record on the internet, etc.

Advantages

No official registration required. Comes into effect immediately. No cost to marking something as copyright.

Disadvantages

Does not protect ideas, only the way in which the idea is expressed To be effective, it must be enforced. You have responsibility for legally asserting copyright, if you feel it is infringed. Involves legal costs and proving infringement.

Patent

Granted by a government, a patent is the right for a limited period to stop others making, using or selling an invention without permission. Patents are concerned with products or process which are:

New, cannot have been made public in anyway before the date the patent is filed Involve an inventive step. Be capable of industrial application, as distinct from a purely intellectual or aesthetic activity. Process alleged to operate in a manner contrary to well-established physical laws, therefore, do not have industrial application. Patents can be taken on a country specific basis or worldwide. More countries mean more cost and effort to achieve this.

Advantages

A patent gives the legal right to stop others using your invention. Its existence may be enough to deter competitors. Buys time (20 years) in which an inventor can develop a market to the product or process.

Attractive to investors, as it limits competition.

Disadvantages

Takes time and money to establish a patent. All patents have to be 'researched' to ensure there are no existing patents of a similar nature - involves legal fees. Not possible to guarantee that a patent (once granted) is valid, it can be legally challenged and revoked with no refunds It is still upto the inventor to protect a patent, if they can identify an infringement - the patent office don't take sides.

Granting a patent is no indication that the invention has any merit or commercial value. Some products and process can be varied slightly to get around the exact wording of Patents.

Industrial Design

Monopoly right for the appearance of the whole or a part of a product resulting from the features lines, colours, shape, texture or materials used. Cannot be the same, as any design already available and must have 'individual characters.

Advantages

Can be combined with legal protection from copyright and design rights. Protect genuine design innovation for upto 25 years.

Disadvantages

Cannot be applied to designs that concern how a product works or which are not visible in normal use - ascetics only. Easily subverted by slight amendments to the basic design.

Annexures

Annexure-I

Modal Questions

1. What do you mean by Intellectual Property?
2. What are "intellectual property rights?
3. Why does intellectual property need to be promoted and protected?
4. What are the legislations covering IPRs in India?
5. Who are responsible for administration of IPRs in the country?
6. India provides protection to Intellectual Property Rights in accordance with its obligations under the TRIPS Agreement of the WTO. Discuss.
7. List out the subject-matter protected by intellectual property rights under the World Intellectual Property Organization (WIPO).
8. Enumerate the salient features of the Design Act, 2000.
9. Outline the main differences between utility models and patents. Does India have legislation on Utility models?
10. What is Convention on Biological Diversity (CBD)? Is India a member of CBD? If yes, explain the approach adopted by India towards Biological Diversity.
11. What is the legal position in India for protecting New Plant Varieties?
12. What are the objects and purposes behind plant varieties protection law?
13. Explain the competing rationales for protection of IPRs.
14. What is TRIPS Agreement? Outline the main three features of the TRIPS Agreement.

15. Summarize the provisions of Paris Convention for the Protection of Industrial Property.
16. What is the World Intellectual Property Organization?
17. How does WIPO promote the protection of intellectual property?
18. Write short notes on the following:
    i) Utility Models
    ii) The trademarks
    iii) Copyright
    iv) Geographical Indication

1. What are Intellectual Property Rights (IPR)?

Ans: IPR is a general term covering patents, copyright, trademark, industrial designs, geographical indications, layout design of integrated circuits, undisclosed information (trade secrets) and new plant varieties.

2. What are the legislations covering IPRs in India?

Ans:

- Patents: The Patents Act, 1970 as amended in 1999, 2002 and 2005 Design:
- The Designs Act, 2000
- Trademark: The Trademarks Act, 1999
- Copyright: The Copyright Act, 1957 as amended in 1983, 1984 and 1992, 1994, 1999 Layout Design of Integrated Circuits
- The Semi-conductor Integrated Circuits Layout Design Act, 2000
- Protection of Undisclosed Information: No exclusive legislation exists, but the matter would be generally covered under the Contract Act, 1872
- Geographical Indications: The Geographical Indications of Goods (Registration and Protection) Act, 1999
- Plant Varieties: The Protection of Plant Variety and Farmers' Rights Act, 2013.

3. Who are responsible for administration of IPRs in the country?

Ans: Patents, designs, trademarks and geographical indications are administered by the Controller General of Patents, Designs and Trademarks which is under the control of the Department of Industrial Policy and Promotion, Ministry of Commerce and Industry. Copyright is under the charge of the Ministry of Human Resource Development. The Act on Layout-Design of Integrated Circuits is administered by the Ministry of Telecommunication and Information Technology. Protection of Plant Varieties and Farmers' Rights Authority, Ministry of Agriculture administers the Act on Plant Variety.

4. Give your Answer

Patent

COPYRIGHT EXAMPLES

Drawings and writing

Geographical Indication (GI)

Design

20. Fill in the Blanks

i) **Intellectual property** (IP) refers to any original creation of the human intellect, such as artistic, literary, technical, or scientific creation.

ii) **Intellectual property rights** (IPR) refers to the legal rights given to the inventor or creator to protect his invention or creation for a certain period of time.

iii) The first Letter Patent granted for an invention took place in the year **1449.**

iv) France set the first legal system to protect trade marks in world in the year **1857.**

v) Trademark laws first introduced in Britain in the year **1862.**

vi) Importance of IP first recognized in the Paris Convention in **1883** for the Protection of Industrial Property.

vii) One of the most important modern developments in international intellectual property protection is the establishment of the World Intellectual Property Organisation (WIPO) in the year **1967.**

viii) The Patents Act (Act 39 of 1970) came into force on **20th April, 1972.**

ix) Patents have a term of **20 years** from the date of filing a complete specification.

x) **Copyright** is valid for the lifetime of the author and minimum 50 years after the death of the author.

xi) **Geographical Indication** term is initially for a period of 10 years and can be renewed perpetually.

xii) The Patent Law Treaty (PLT), adopted by WIPO **in June of 2000,** entered into force on **April 28, 2005**.

xiii) The **TRIPS** Agreement came into force in 1995, as part of the Agreement Establishing the World Trade Organization.

xiv) **Copyright** is a legal term used to describe the rights that creators have over their literary and artistic works.

xv) Darjeeling tea is the example of **geographical indication.**

xvi) In Rajasthan- Bikaneri Bhujia is the example of **geographical indication.**

xvii) In Andhra Pradesh- Tirupati Laddu is the example of **geographical indication**

xviii) In Maharashtra- Mahabaleshwar strawberry is the example of **geographical indication.**

xix) The Plant Variety Protection and Farmers Rights Act, 2001 has come into force in **2005**.

xx) Joha rice is the example of geographical indication for the state **Assam.**

Important Abbreviations and Authorities In IPR

IP	Intellectual Property
IPR	Intellectual Property Right
GATT	General Agreement of Tariffs and Trade
SPS	Sanitary and Phyto-Sanitary measures
TBT	Technical Barriers to Trade
TRIPS	Trade Related aspects of Intellectual Property Rights
WTO	World Trade Organization
UPOV	International Union for the Protection of New Varieties of Plants
WIPO	World Intellectual Property Organization
ITPGRFA	International Treaty on Plant Genetic Resources for Food and Agriculture
PPV&FRA	Protection of Plant Varieties & Farmers' Rights Authority
NBA	National Biodiversity Authority
PCT	Patent Co-operation Treaty

Annexure-II

Official Address

1. Indian Patent Office/Trademarks Registry/Designs Office
   Central Government Office
   Kolkata, West Bengal
   Ph. 098998 01721
2. Rajiv Gandhi National Institute of Intellectual
   Property Management
   (18) Government College
   Nagpur, Maharashtra
   0712 254 0913
3. Intellectual Property India
   (71) Patent Office
   New Delhi, Delhi
   011 2803 2491
4. The Patent Office
   County Government Office
   Chennai, Tamil Nadu
   044 2250 2039
5. Intellectual Property India
   Government Office
   Chennai, Tamil Nadu
   044 2250 2081
6. Patent in India Pune office
   Patent Attorney
   Pune, Maharashtra
   080555 63001
7. Indian Patent and Trademarks Office
   Central Government Office
   Mumbai, Maharashtra
   022 2413 2735
8. InvnTree Intellectual Property Services Pvt. Ltd.
   Patent Attorney

Bengaluru, Karnataka
080 3946 7555

9. Intellectual Property Bhawan
Government Office
Kolkata, West Bengal
033 2367 9101

10. SSRANA & Co | Patent Trademark Copyright Design Filing in India
Law Firm
New Delhi, Delhi
011 4012 3000

11. Patent Information System
No reviews · Information Bureau
Nagpur, Maharashtra
0712 251 1241

12. Patent in India Platform Mumbai
Patent Attorney
Greater Mumbai, Maharashtra
080555 63001

13. Intepat IP Services Pvt Ltd
Patent Attorney
Bengaluru, Karnataka
080 4217 3649

14. Patent Attorney
Ahmedabad, Gujarat
093746 56660

15. Boudhik Sampada Bhavan, Trademarks & Patent Office, Kolkata
Government Office
Kolkata, West Bengal
098998 01721

16. Krislaw (Pune Trademark Lawyers, Trademark List, Indian Trademark Association)

No reviews · Patent Attorney
Pune, Maharashtra
020 4108 5501

17. ATI IPR Firm - Trademark Application Firm in Pune, Indian Trademark Attorney
No reviews · Patent Attorney
Pune, Maharashtra

Annexure-III

"**FORM 1** THE PATENTS ACT 1970 (39 of 1970) and THE PATENTS RULES, 2003 **APPLICATION FOR GRANTOFPATENT** (See section 7, 54 and 135 and sub-rule (1) of rule 20)	(FOR OFFICE USE ONLY)	
	Application No.	
	Filing date:	
	Amount of Fee paid:	
	CBR No:	
	Signature:	
1.APPLICANT'S REFERENCE / IDENTIFICATION NO. (AS ALLOTTED BY OFFICE)		

2. TYPE OF APPLICATION [Please tick (✍) at the appropriate category]					
Ordinary ( )		Convention ( )		PCT-NP ( )	
Divisional ( )	Patent of Addition ( )	Divisional ( )	Patent of Addition ( )	Divisional ( )	Patent of Addition ( )

3A. APPLICANT(S)				
Name in Full	Nationality	Country of Residence	Address of the Applicant	
			House No.	
			Street	
			City	
			State	
			Country	
			Pin code	

3B. CATEGORY OF APPLICANT [Please tick (✍) at the appropriate category]			
Natural Person ( )	Other than Natural Person		
	Small Entity ( )	Startup ( )	Others ( )

4. INVENTOR(S) [Please tick (✍) at the appropriate category]		
Are all the inventor(s)	Yes ( )	No ( )

same as the applicant(s)
named above?

If "No", furnish the details of the inventor(s)				
Name in Full	Nationality	Country of Residence	Address of the Inventor	
			House No.	
			Street	
			City	
			State	
			Country	
			Pin code	

5. TITLE OF THE INVENTION

6. AUTHORISED REGISTERED PATENT AGENT(S)	IN/PA No.	
	Name	
	Mobile No.	
7. ADDRESS FOR SERVICE OF APPLICANT IN INDIA	Name	
	Postal Address	
	Telephone No.	
	Mobile No.	
	Fax No.	
	E-mail ID	

8. INCASE OF APPLICATION CLAIMING PRIORITY OF APPLICATION FILED IN CONVENTION COUNTRY, PARTICULARS OF CONVENTION APPLICATION					
Country	Application Number	Filing date	Name of the applicant	Title of the invention	IPC (as classified in the convention country)

9. IN CASE OF PCT NATIONAL PHASE APPLICATION, PARTICULARS OF INTERNATIONAL APPLICATION FILED UNDER PATENT CO-OPERATION TREATY (PCT)	
International application number	International filing date

10. IN CASE OF DIVISIONAL APPLICATION FILED UNDER SECTION 16, PARTICULARS OF ORIGINAL (FIRST) APPLICATION	
Original (first) application No.	Date of filing of original (first) application

11. IN CASE OF PATENT OF ADDITION FILED UNDER SECTION 54, PARTICULARS OF MAIN APPLICATION OR PATENT	
Main application/patent No.	Date of filing of main application
12. DECLARATIONS	
(i) Declaration by the inventor(s) (**In case the applicant is an assignee**: the inventor(s) may sign here in below or the applicant may upload the assignment or enclose the assignment with this application for patent or send the assignment by post/electronic transmission duly authenticated within the prescribedperiod). I/We, the above named inventor(s) is/are the true & first inventor(s) for this Invention and declare that the applicant(s) herein is/are my/our assignee or legal representative. (a)Date (b)Signature(s) (c)Name(s)	
(ii) Declaration by the applicant(s) in the convention country (In casethe applicant in India is different thanthe applicant in theconvention country: the applicant in the convention country may sign herein below or applicant in India may upload the assignment from the applicant in the convention country or en close the said assignment with this application for patent or send the assignment by post/electronic transmission duly authenticated within the prescribed period) I/We, the applicant(s) in the convention country declare that the applicant(s) herein is/are my/our assignee or legal representative. (a)Date (b)Signature(s) (c)Name(s) of thesignatory	

(iii) Declaratio by the applicant(s)

I/We the applicant(s) hereby declare(s) that: -

- I am/ We are in possession of the above-mentioned invention.
- The provisional/complete specification relating to the invention is filed with this application.
- The invention as disclosed in the specification uses the biological material from India and the necessary permission from the competent authority shall be submitted by me/us before the grant of patent to me/us.
- There is no lawful ground of objection(s) to the grant of the Patent to me/us.
- I am/we are the true & first inventor(s).
- I am/we are the assignee or legal representative of true & first inventor(s).
- The application or each of the applications, particulars of which are given in Paragraph-8, was the first application in convention country/countries in respect of my/our invention(s).
- I/We claim the priority from the above mentioned application(s) filed in convention country/countries and state that no application for protection in respect of the invention had been made in a convention country before that date by me/us or by any person from which I/We derive the title.
- My/our application in India is based on international application un der Patent Cooperation Treaty (PCT) as mentioned in Paragraph-9.
- The application is divided out of my /our application particulars of which is given in Paragraph-10 and pray that this application may be treated as deemed to have been filed on DD/MM/YYYY under section 16 of the Act.
- The said invention is an improvement in or modification of the invention particulars of which are given in Paragraph-11.

13. FOLLOWING ARE THE ATTACHMENTS WITH THE APPLICATION

(a) Form 2

Item	Details	Fee	Remarks
Complete/ provisional specification)#	No. of pages		
No. of Claim(s)	No. of claims and No. of pages		
Abstract	No. of pages		
No. of Drawing(s)	No. of drawings and No. of pages		

In case of a complete specification, if the applicant desires to adopt the drawings filed with his provisional specification as the drawings or part of the drawings for the complete specification under rule 13(4), the number of such pages filed with the provisional specification are required to be mentioned here.

(b) Complete specification (in conformation with the international application)/as amended before the International Preliminary Examination Authority (IPEA), as applicable (2 copies).

(c) Sequence listing in electronic form

(d) Drawings (in conformation with the international application)/as amended before the International Preliminary Examination Authority (IPEA), as applicable (2 copies).

(e) Priority document(s) or a request to retrieve the priority document(s) from DAS (Digital Access Service) if the applicant had already requested the office of first filing to make the priority document(s) available to DAS.

(f) Translation of priority document/ Specification/ International Search Report/ International Preliminary Report on Patentability.

(g) Statement and Undertaking on Form 3

(h) Declaration of Inventorship on Form 5

(i) Power of Authority

(j)..

Total fee ?..........in Cash/Banker's Cheque/Bank Draft bearing No.......... Date...............on **.............Bank.**

I/We hereby declare that to the best of my/our knowledge, information and belief the fact and matters slated herein are correct and I/We request that a patent may be granted to me/us for the said invention.

Dated this.....................day of......... 20........................

Signature:

Name:

To,

The Controller of Patents

The Patent Office, at...................

Note: -

* Repeat boxes in case of more than one entry.
* To be signed by the applicant(s) or by authorized registered patent agent otherwise where mentioned.
* Tick (✍)/cross (x) whichever is applicable/not applicable in declaration in paragraph-12.
* Name of the inventor and applicant should be given in full, family name in the beginning.
* Strike out the portion which is/are not applicable.
* For fee: See First Schedule";

❑❑❑

References

1. A.S. Holzwarth. 1998. Patent Protection and Computer Programs in the Practice of the European Patent Office, WIPO/CNR/KYI/98/3.
2. Andrew Christie. 1995. Integrated Circuits and their Contents: International Protection, Ed. Sweet & Maxwell, London, UK.
3. Brito, Jerry. 2012. Eds, Copyright Unbalanced: From Incentive to Excess, Mercatus Center at George Mason University: Arlington
4. Drahos, Peter. 1996. A Philosophy of Intellectual Property, Dartmouth: Aldershot.
5. Flew, Terry, Leisten, Susanna H., & Hearn, Gregory N. 2006. Alternative syst ems for Intellectual Property in the digital age. In: Servaes, Jan & Thomas, Pradip (Eds.) Intellectual Property Rights Communication in Asia: Conflicting Traditions. SAGE Publications: New Delhi, pp. 226 240.
6. Horten, Monica. 2013. A Copyright Masquerade, Zed Books: London http://www.dcmsme.gov.in/schemes/Guidelines-UK.pdf
7. Hughes, Justin. 1988. The Philosophy of Intellectual Property, 77 Geo. L.J. 287, Retreioved from http://www.justinhughes.net/docs/a-ip01.
8. Intellectual property (TRIPS) - Agreement text - standards. *WTO. 1994-04-15*. Retrieved 2012-04-16.
9. International Bureau of WIPO, Basic Facts about the Patent Cooperation Treaty, WIPO Pub. No. 433.
10. International Bureau of WIPO, Budapest Treaty on the International Recognition of the Deposit of Microorganisms for the Purposes of Patent Procedure and Regulations 1977, as amended upto January 1981, WIPO Pub. No. 277

11. International Bureau of WIPO, Character Merchandising, WO/INF/108, December 1994
12. International Bureau of WIPO, General Information on the Seventh Edition of the International Patent Classification, WIPO Pub. No. 409, 1998
13. International Bureau of WIPO, Guide to the Deposit of Microorganisms under the Budapest Treaty, WIPO Pub. No. 661.
14. International Bureau of WIPO, Information on the Budapest Treaty on the International Recognition of the Deposit of Microorganisms for the Purposes of Patent Procedure, WIPO/IP/MOW/00/20, October 2000
15. International Bureau of WIPO, Introduction to Trademark Law & Practice (second edition), WIPO Pub. No. 653, 1993
16. International Bureau of WIPO, IPC:CLASS for Windows—The International Patent Classification (IPC) Search System, 1999
17. International Bureau of WIPO, Madrid Agreement Concerning the International Registration of Marks; Protocol Relating to the Madrid Agreement Concerning the International Registration of Marks; Regulations, WIPO Pub. No. 204, 2004
18. International Bureau of WIPO, Paris Convention for the Protection of Industrial Property 1883, as amended up to September 1979, WIPO Pub. No. 201
19. International Bureau of WIPO, Patent Cooperation Treaty (PCT) and Regulations under the PCT, WIPO Pub. No. 274, 2004
20. International Bureau of WIPO, Patent Law Treaty (PLT), and Regulations under the Patent Law
21. International Bureau of WIPO, PCT Applicant's Guide, WIPO Pub. No. 432.
22. International Bureau of WIPO, PCT Gazette, WIPO Pub. No. 108 (published weekly) Home page: http://www.wipo.int/pct/en/index.html
23. International Bureau of WIPO, PCT Newsletter, WIPO Pub. No. 115 (published monthly)
24. International Bureau of WIPO, Records of the Washington Diplomatic Conference on the Patent Co-operation Treaty, WIPO Pub. No. 313, 1972

25. International Bureau of WIPO, Staff Training and Development: Use of Patent Information Systems, WIPO/R&D/SIN/97/6, October 1997

26. International Bureau of WIPO, The First Twenty-Five Years of the Patent Co-operation Treaty (PCT) (1970-1995), WIPO Pub. No. 884, 1995

27. J. Laddie, Issues Arising in Litigation in the Field of Layout-Designs of Integrated Circuits and Unfair Competition (Including the Misuse of Confidential Information) - Outline, WIPO/IP/POS/98/6,November 1998

28. John Frow. 2006. Intellectual property rights and the Public domain in the new world order.

29. Lemley, M.A., Shapiro, C. 2005. "Probabilistic patents". The Journal of Economic Perspectives. 19 (2): 75–98

30. M. Blakeney, Trademarks and Passing Off, WIPO/IPLT/ABV/99/9, September 1999

31. Magnus, David. 2002. "Intellectual Property and Agriculture Biotechnology: Bioprospecting or biopiracy?"Amherst NY: Prometheus Books. 265-276

32. Murphy, D.J. 2012. Are intellectual property rights compatible with Rawlsian principles of justice?Ethics and Information Technology, 14 (2), pp.109-121.

33. P. Drahos, Patents, Designs and Utility Models, WIPO/IPLT/ABV/99/8, September 1999

34. P. Nuss, Franchising, WIPO/CEIPI/IP/SB/00/15, September 2000

35. Rajkumar S. Adukia.Handbook on intellectual property rights in India. www.caaa.in/image/34_hb_on_ipr.pdf

36. Shiva, Vandana. 1997. Biopiracy: The Plunder of Nature and Knowledge. Cambridge MA: South End Press

37. The Indian journal of law and technology, 2: 106-126

38. Wang, Chunyan. 2008. Creative Commons License: An Alternative Solution to Copyright in the New Media Arena, In: Fitzgerald, B., Gao, F., O'Brien, D, &Xiaoxiang Shi, S. (Eds), Copyright Law, Digital Content

39. What is the PLT (Patent Law Treaty)? WIPO Pub. No. L450PLT, 2000 (leaflet)

40. WIPO Intellectual Property Handbook: Policy, Law and Use

41. Wolf, Clark. 2004. "Plants, Patents. Property, and Pirates Part I: Luther Burbank versus Thomas Jefferson" Iowa State University: Bioethics in Brief.

42. World Trade Organization, "Part I – General Provisions and Basic Principles", *Agreement on Trade-Related Aspects of Intellectual Property Rights*

43. World Trade Organization, "Part II – Standards concerning the availability, scope and use of Intellectual Property Rights; Sections 5 and 6", *Agreement on Trade-Related Aspects of Intellectual Property Rights*

□□□

About the Authors

Kalyan Sarma was born on 1st March, 1970 in Kamrup district of Assam. He has earned B.V.Sc. & A.H. degree from Assam Agricultural University, Jorhat. He has completed his M.V.Sc. degree with distinction in the year 1996 from Assam Agricultural University, Jorhat in the discipline Veterinary Medicine and Ph.D. degree from Indian Veterinary Research Institute, Bareilly in the year 2012. He has published three books, above 100 research articles at different national and international journal, three manuals and several popular articles and folder. He has attended many national and international seminars, symposium and refresher training courses. He has received "Overseas Associate Fellowship". Now- a-days he is working as Associate Professor in Department of Veterinary Medicine and actively involved in teaching, research and extension activities at College of Veterinary Sciences & Animal Husbandry, Central Agricultural University, Selesih, Aizawl, (Mizoram) India.

Dibyajyoti Talukdar was born on 22nd July, 1983 in Kamrup district of Assam. He has earned B.V.Sc. & A.H. degree from Assam Agricultural University, Jorhat. He has completed his M.V.Sc. degree with distinction in the year 2010 from Assam Agricultural University, Jorhat in the discipline Animal Reproduction, Gynaecology & Obstetrics and Ph.D. degree in the year 2014. Presently, he is working as Assistant Professor at Department of Animal Reproduction, Gynaecology & Obstetrics, College of Veterinary Sciences & Animal Husbandry, Central Agricultural University, Selesih, Aizawl, Mizoram, India. He has published two books, thirty eight research articles at different national and international journal, four manuals and several popular articles and folder. He has attended many national and international seminars, symposium and refresher training courses. Now- a-days he is actively involved in teaching, research and extension activities at College of Veterinary Sciences & Animal Husbandry, Central Agricultural University, Selesih, Aizawl, (Mizoram) India.